Science Culture, Language, and Education in America

Emily Schoerning

Science Culture, Language, and Education in America

Literacy, Conflict, and Successful Outreach

palgrave
macmillan

Emily Schoerning
University of Iowa
Iowa City, IA, USA

ISBN 978-1-349-95812-2   ISBN 978-1-349-95813-9   (eBook)
https://doi.org/10.1057/978-1-349-95813-9

Library of Congress Control Number: 2018942018

Cover Photo © Blend Images / Alamy Stock Photo

Printed on acid-free paper

This Palgrave Macmillan imprint is published by the registered company Nature America,
Inc. part of Springer Nature.
The registered company address is: 1 New York Plaza, New York, NY 10004, U.S.A.

CONTENTS

CHAPTER 1

Where Are We Now? Where Could We Be?

Imagine if, here in America, you could send your kids to public school and know what kind of science education they would experience. Not a series of disconnected science facts and equations, but instead an approach to life that employs scientific thinking as part of a toolkit of skills for navigating the world. And along with these practices, confidence: a sense that they know what it means to think and act as a scientist.

Imagine what it would be like if, as these kids grew up, they were exposed to a world of science that looked a lot like their world. A scientific community peopled by scientists who looked like them and sounded like them, men and women from all sorts of backgrounds. A scientific community composed of the same diversity of people we find in our nation.

This thought exercise is surprisingly difficult. Our current American reality is so far off from what it could be, both in terms of how science is often taught in the classroom and how inaccurately the composition of Science, Technology, Engineering, and Math (STEM) professionals reflects American diversity. This is not to say that our current reality is all bad. After all, the United States has the most vibrant scientific research sector in the world.[1] American research is consistently held up as extraordinarily creative and innovative: an acknowledged power behind the lasting economic strength of the United States.[2] And yet, despite years of effort to promote diversity, the science workforce remains overwhelmingly white and male, especially at its highest levels.[3] Girls, children of lower socioeconomic status, and children of color either do not pursue or do not complete

© The Author(s) 2018
E. Schoerning, *Science Culture, Language, and Education in America*, https://doi.org/10.1057/978-1-349-95813-9_1

training in the sciences. At the same time, approximately one-third of Americans reject the theory of evolution, nearly half do not believe that human activities are responsible for climate change, and over sixty percent believe that genetically modified foods are not safe to eat.[4] A small but significant minority of parents believe that vaccines are unsafe. In each case, huge numbers of adults are rejecting findings that are overwhelmingly accepted by the scientific community.[5]

This indicates either a substantial, community-wide problem with science education and resulting adult scientific literacy or a significant cultural disconnect, or both. A good deal of critical attention is being paid to potential problems in science education. Both research and investment in STEM education have grown steadily in the past decade. In some cases, advocacy toward STEM investment simply means urging or requiring students to take more science courses in order to improve their science literacy. In the best scenarios, new STEM initiatives focus on experiential, hands-on approaches that introduce students to "the practice of science". The idea here, supported by substantial research-based evidence, is that if students actually experience what it means to do science, if they have a real encounter with the culture of science, they will be more likely to pursue STEM careers. The mantra: teach more science, and teach it better. In many circles, this is assumed to be enough both to attract more underrepresented groups into the sciences and to improve the science literacy of the population overall.

But what if the culture of science is itself part of the problem? In this book, when I speak of the culture of science, I refer broadly to the interactions between the language of science, conceptions regarding the purpose and nature of science (NOS), and the various media, spaces, and forms through which the scientific community and the general public communicate.[6] While classroom science practices do receive a fair amount of critical attention and research, this second potential issue, that science as it is currently presented and practiced may offer significant cultural conflicts for many Americans, receives less attention. The topic is less widely researched, and to a significant degree less understood as an area that could benefit substantially from research. If we work to bring more science into the classroom, to bring students up close and personal with science, what do we do if this is not enough to solve the problem? What if students find science not to be exciting and appealing but unfriendly, intimidating, and foreign to their experience?

The two troubling realities—the American scientific community that lacks diversity and the American population that has a tendency to reject scientific findings—are not unrelated. Indeed, they both stem in part from the same root cause. The culture of science, as portrayed in the media, and as transmitted in science classes at all levels, is perceived by many Americans as alienating and exclusionary.

If the culture of science is exclusionary, we have a problem, because we must also acknowledge that this culture has power. The ability to engage with STEM offers individuals personal, economic, and political power.[7] Engaging with STEM is relevant to people's job prospects and financial security, as well as their ability to make informed personal and political decisions. Finding ways to help individual Americans develop a sense of access and power related to the culture of science is important for the enfranchisement of all Americans.[8]

Before we examine science as a culture, it is worth considering its inherent power as a discipline of thought. In the discipline of science, one acquires both content knowledge and a variety of skills related to questioning, research, and analysis. Possessing knowledge alone imbues one with power, but the practical skills associated with science give people the even more powerful abilities to produce knowledge and answer questions.[9] These are both extremely potent tools that individuals can make use of in their daily lives.[10] Individuals who are comfortable researching and evaluating information about their medical care, who can understand and interpret the numbers on nutritional labels, and who can read manuals and use tools to do home repair are exercising the power of STEM in their personal lives. These behaviors increase people's everyday health and well-being.

Beyond these personal agentic concerns, the power of science as a discipline has, at this time in our nation's history, a great deal of concrete economic value. As automation and off-shoring reduce the quantity of good jobs in the United States, people rely more and more on their educations to find secure employment and financial stability. Many new jobs and opportunities are being created in the STEM fields, which include a diverse spectrum of specialties from healthcare to tech start-ups. Employment in the STEM fields is generally higher paying, more secure, and in many ways more pleasant than other available American jobs, particularly as we move increasingly in the direction of a service economy. Furthermore, STEM-based industries that exploit local resources from agricultural waste to geothermal energy to wind power can provide opportunities to agricultural, manufacturing, and other communities fighting to adapt to changing economic conditions. Given these potent realities, why are relatively few young people, especially from marginalized groups, pursuing scientific careers? Could it be because they simply cannot envision themselves as scientists[11]?

Here, we begin to examine what impact the culture of science has on the American population. Why do people, most of whom are aware of the personal and economic power related to the practice of science as a discipline, reject science for themselves? To put it more simply, why don't more students go into STEM, and why do they fail to see themselves as scientists?

This is a more profound question than simply "when they look at scientists do they see people that look like them?" Partly because of how science is presented in many American classrooms, many students do not come to see themselves as potential participants in the scientific enterprise.[12] The vast majority of people, regardless of their ethnic background, gender, or economic status, look at science as a discipline and a community, and don't think of themselves as belonging there. They don't see a possibility of being involved in scientific discourse. Science, like so many other topics in school, is often perceived by students as simply something to take in, not a process with which to engage. Significantly, the way in which students are expected to internalize science in our culture almost always starts with learning the specialized language of science. In our school system, the first introduction to science as an experience, as a culture, does not come through the practice of science but through the language of science. This language is highly specialized and is often extremely disconnected from students' home language practices, both in its vocabulary and discourses practices.[13]

Before going on to describe some of the problems with the way science is often conveyed in the classroom, and make concrete and practical suggestions for improvement, it is crucial to emphasize that blame for the current situation should not be placed on science teachers. Teachers should not be blamed for conveying the culture of science as they themselves received it, and which they embraced when they took up the challenge of conveying science to the next generation. Nor should the reader get the impression that we are suggesting that there is, somehow, something intentionally or unchangeably wrong with the culture of science. Within the scientific community, the accepted methods and styles of discourse and debate are effective and efficient. What we aim to achieve is heightened awareness of the disconnect between how scientists talk and how teachers portray science in the classroom, and how that style of discourse can be deeply, and sometimes permanently, alienating.

Both traditional and several current common methods of science teaching do not help the majority of students develop a sense of access and power related to the discipline of science, because it is assumed that learning that language must come first. The language of science is notable for its precision and accuracy. Imagine if you took a small child, say two years old, and showed it a dog. Most people would encourage the child in what description the child could give of the dog; if the child said "a dog!" most people would react positively, and encourage further description. This is not how language is taught in science class. Under the science classroom model of language acquisition, if the child could not identify the dog as, say, a wire-haired terrier, it is unlikely they would earn any praise, and would more likely be subject to correction. In many science classrooms, with many well-meaning teachers, and with quite young children, I have seen children corrected rather than encouraged, and silenced for incorrect responses.

For example, during observations in middle-American science classrooms, third graders learning about the rock cycle had their questions ignored and their statements dismissed if they did not use appropriate scientific vocabulary.[14] Teachers looked for vocabulary words. Students who used phrases that showed understanding, for example "rocks that are squished together!" as related to sedimentary rocks, were told they were wrong. During classroom observations I have often seen that students who mispronounced words were teased by their peers and, in some sad cases, by their teachers.

These trends often continue throughout students' educations, and create a notable silencing effect. It is difficult for many people to accurately pronounce new multisyllabic words without practice. Students may be unwilling to take the social risk of "talking science" if they fear using these words will cause them social embarrassment, a fear held very keenly by most children. Students who are literally rendered mute by the language of science often develop a predictable and understandable disengagement from science as a discipline. Even those students who do become fluent in the language of science may be weaker in their conceptual grounding than they should be, because mastery of vocabulary is too often equated with understanding in American science classrooms.[15,16]

Students' home language practices may resemble the classroom science language practices to greater or lesser degrees.[17] Students of high socioeconomic status from native English-speaking backgrounds are more likely to be exposed to the type of vocabulary found in the science classroom at home, and are also more likely to be asked open-ended questions by their parents. Students of lower socioeconomic status, even if they are native English speakers, are exposed to substantially less vocabulary in their home language practices, and accordingly are less likely to have practice adding new words to their internal lexicons.[18] They are also less likely to be asked open-ended questions by their parents, and thus less likely to be comfortable with the language practices of science. Students who are non-native English speakers, or who are from cultures that do not traditionally teach children through open-ended questioning, can face additional challenges when they encounter the linguistic practices of science.[19,20] Language and culture are intimately related. Students who have difficulty accessing the language of science are often unable to access or acquire other aspects of the culture of science.[21]

The culture of science, when analyzed through the lens of linguistics, displays many discourse practices that differ from those of the American public in ways that go far beyond vocabulary. The ways in which scientists communicate information to others in the community, reach consensus, and disseminate conclusions to people both within and outside of the community, are different and distinct from those of the general population. These cultural linguistic behaviors, which rely heavily on a form of dialog that is not widely engaged in across the American cultural spectrum, are often not taught explicitly to students during the course of their science education. Even beyond simple acquisition, the larger purpose of the language and discourse patterns used in science is generally not addressed in the science classroom, remaining an additional, significant, and challenging topic that must be inferred by the student. By failing to address issues of access to the language of science, students often become or feel that they are locked out of the culture of science. The perception of being locked out from a culture can, for obvious reasons, lead to disengagement.

The disengagement that results from our current linguistic approach to science education negatively affects many people early in their educations. Their sense of disengagement continues to be significant in other aspects of adult life. Many of today's political issues require a command of scientific concepts and science practice for meaningful understanding.[22] Topics like climate change, evolution, reproductive rights, and stem cell research are all highly politicized. They all require a fairly sophisticated grasp of both scientific concepts and process to understand, and to understand what arguments presented in regard to these issues are not scientific, representing misinformation at best and deliberate propaganda at worst. The decisions we make about all of these issues will have major future impacts on American life.

Disenfranchisement from science leads to political disenfranchisement. A population that is insufficiently able to understand how science works is unable to make informed political choices.

These issues, personal and political disenfranchisement, as well as economic security, are crucial in modern American life, and are deeply tied to STEM practices. Despite their importance, our current American reality, both in terms of how science is taught and who grows up to populate our research and high-technology sectors, is very far off from what it could be. When we send our children to public school today, we know that sixty percent of American teachers do not cover the topic of evolution to state standards. Another thirteen percent actively teach creationism. Evolution, which is central to the understanding of biology, and other potentially socially controversial topics, such as climate change, are often simply avoided despite their importance and their established place in the Next Generation Science Standards (NGSS), which have currently been accepted by nineteen states.[23]

Why is this? We argue this lack of science literacy is causally tied to the difficulty people experience accessing science as a culture. Just as linguistic practices cause many people to feel locked out of the culture of science, so does a lack of authentic science practices in the American science classroom create cultural isolation.[24] The lack of authentic science practice is causally related to the frequent and common science content and science practice misconceptions found throughout the American population.[25] Rather than being taught as a dynamic, evidence-based method of observing the world and answering questions, science is often and inaccurately taught as a monolithic and static body of facts.[26] People are taught to conceptualize science as something that is, rather than something you do. This, a serious misconception in and of itself, leads to many other misconceptions about science and its nature. These misconceptions can be damaging in many ways, as they often lead to the disenfranchisement of individuals connected to the scientific community and scientific practice.

Damaging misconceptions related to science include both process misconceptions and content misconceptions. Process misconceptions, such as the notion that a scientific theory is "just a theory", or the idea that scientific knowledge is ultimate or static, expose global weakness in an individual's understanding of science as a discipline. Content misconceptions, which are also common, may be held by individuals with a reasonably well-founded understanding of the discipline of science, but who have learned or been taught inaccurate information. These types of misconceptions can often be corrected with outreach and evidence, whereas process misconceptions are more difficult, but not impossible, to change.[27]

It is particularly important to note that when science is interpreted as a static set of facts and a monolithic worldview, a common outcome of process misconceptions, it is more likely for science to be seen as engaged in cultural conflict with religion. The American media often portrays science and religion in juxtaposition, as if they are in necessary and unavoidable conflict. However, this is not the case. The majority of religious groups represented in the United States do not experience fundamental conflicts between a scientific and a religious worldview.[28] When surveyed, a strong majority of Americans from diverse faith backgrounds did not think that their religious views were in conflict with evolutionary theory, an issue that is presented in the media as perhaps the great divide between science and religion. Religious leaders from Christian, Muslim, Jewish, and Buddhist communities, among others, openly state that there is no conflict between science and their faiths. The fact that so many religious people do not experience this conflict makes the political reality of the science/religion conflict extremely interesting. Clearly, this conflict does exist for some people and organizations. It is necessary to understand the ways in which misconceptions about science and science culture drive this cultural conflict if we are to practice effective STEM outreach, as well as improve American science education so as to reach all American students.

Though there are many wonderful science teachers in our country, and many well-intentioned educators and administrators, the reality of science education in America is that it often falls short of our goals. Although national standards increasingly call for an education grounded in science practice and questioning, an education that includes more authentic and accessible exposure to the culture of science, in many classrooms science education remains very traditional, failing to meet standards regarding argumentation.[29,30] Students experience cookbook labs at best, with little opportunity to engage in actual questioning or discovery. Simple memorization remains a standard teaching method in many schools, with students learning and being tested on long lists of science vocabulary words. This utilizes language as a barrier to culture, rather than allowing the development of scientific language in hand with scientific practice.

We have a variety of challenges with science and America that are feeding each other, forming a negative-feedback loop. Students that go into science are demographically disproportionate due to complex issues of culture and power, scientists self-select further due to similar issues of culture and power, and we end up with a demographically disproportionate population of scientists, who then struggle to attract a more diverse group of students to become the next generation of scientists.

The clear interest and real investment the scientific community has toward increasing its diversity, and the progress it has made, should be abiding sources of hope regarding these socially significant issues. If we want to increase diversity in science, we need to address these power issues that impact science education from the very beginning. By addressing these issues in science education, we will be able to improve all students' access to science as a discipline, impacting the lives of not just future scientists, but all Americans.

To change the culture of science, we need to talk about science as a culture. We also need to look critically at the relationships between the culture of science as science is practiced and various other American cultural elements. Over the next few chapters, we'll explore these topics, and present a narrative case study of a program our organization has developed to attempt to address these complex challenges.

First, I'll describe one important aspect of the culture of science, using the lens of language. We'll learn about distinctive cultural features of science as science is practiced, how these differ from classroom science, and what sort of linguistic tools can be introduced to the science classroom to increase student participation and science authenticity, with a focus on dialog.

Misconceptions relating to dialog and science drive a great deal of the conflict relating to science in America; we'll next explore common misconceptions regarding the NOS, and further discuss how these NOS misconceptions relate to societal conflict regarding sources of knowledge.

After looking into broad NOS misconceptions, I'll describe special challenges related to misconceptions about socially controversial topics like climate change and evolution. When traditional teaching methods that ignore the alienating potential of the culture of science are applied to these sensitive topics, outcomes can be particularly unsuccessful due to topical relationships with personal identity issues.

We'll also explore ways religion does and does not contribute to controversy around science in America, where conflict around religion and science often originates, and the diverse ways various world religions understand, include, or reconcile scientific information and the scientific outlook with their teachings, theology, or practices as appropriate.

Then finally, in the last third of the book, I'll describe in some narrative depth my experience developing, piloting, and expanding a program that addresses these cultural issues of science in a holistic fashion, and how the evidence the National Center for Science Education (NCSE) has gathered on the ability of communities to rally around controverisal issues has changed the cultural conversation regarding science.

By the end of this book, I hope you will have had the opportunity to reflect on the complex ways the culture of science interacts with American culture, and the immense possibilities we encounter related to reconciling these cultural differences. We'll begin with a detailed technical look at language, build through applications of language and concept, and end with narrative community experiences. Through this arc, we can come to imagine bigger and broader ideas around science and society. There are ways we can work together to develop genuinely engaging and inclusive science classrooms that serve the diverse American populace, and a science outreach culture that serves our diverse American communities. I hope the ideas you encounter here will help provide some inspiration, and that you will engage in the growing conversation about science in America, as it relates to culture, conflict, and successful outreach.

NOTES

1. Holdren, J. (2013) America COMPETES: Science and the U.S. Economy, S.HRG. 113–641.
2. Kintisch, E. (2005) U.S. economy. Panel calls for more science funding to preserve U.S. prestige. Science 310(5747): 423.
3. National Academy of Engineering. (2014) Advancing Diversity in the US Industrial Science and Engineering Workforce: Summary of a workshop. National Academies Press, Washington DC.
4. Branch, G. (2015) Views on evolution amongst the public and scientists. http://ncse.com/news/2015/01/views-evolution-among-public-scientists-0016160. Accessed 3/15/2017.
5. Branch, G. (2014) Polling confidence in science. http://ncse.com/news/2014/04/polling-confidence-science-0015543. Accessed 3/15/2017.

6. Gee, J. P. (1990) Social linguistics and literacies: Ideology in discourses (2nd ed.). London: Falmer.
7. Hodson, D. (1999) Going beyond cultural pluralism: Science education for sociopolitical action. Science Education 83(6): 775–796.
8. Lewis, C., Enciso, P., & Moje, E. B. (2007) Reframing sociocultural research on literacy: Identity, agency, and power. Mahwah, NJ: Erlbaum.
9. Schoerning, E., & Hand, B. (2013) The discourse of argumentation. Mevlana International Journal of Education 2(3): 43–54.
10. Tippett, C. (2009) Argumentation: The language of science. Journal of Elementary Science Education 21(1): 17–25.
11. Diaz-Rico, L. T., & Weed, K. Z. (2002) The crosscultural, language, and academic development handbook. London: Allyn & Bacon.
12. Hildebrand, G. M. (2001) Re/writing science from the margins. In A. C. Barton & M. D. Osborne (Eds.), Teaching science in diverse settings: Marginalized discourses and classroom practice: pp. 161–199. New York.
13. Lemke, J. (1990) Talking science: Language, learning and values. Norwood, NJ: Ablex.
14. Schoerning, E., et al. (2015) Language, access, and power in the elementary science classroom. Science Education 99(2): 238–259.
15. Richter, E. (2011) The effect of vocabulary on introductory microbiology instruction. Tempe: Arizona State University Press.
16. Schoerning, E. (2014) The effect of plain-English vocabulary on student achievement and classroom culture in college science instruction. International Journal of Science and Mathematics Education 12: 307–327.
17. Choi, I., Nisbett, R. E., & Smith, E. E. (1997) Culture, category salience, and inductive reasoning. Cognition 65(1): 15–32.
18. Lee, O., & Fradd, S. H. (1996) Literacy skills in science learning among linguistically diverse students. Science Education 80(6): 651–671.
19. Duran, B. J. (1998) Language minority students in high school: The role of language in learning biology concepts. Science Education 82(3): 311–341.
20. Rakow, S. J., & Bermudez, A. B. (1993) Science is "Ciencia": Meeting the needs of Hispanic American students. Science Education 77(6): 669–683.
21. Delpit, L., & Dowdy, J. K. (2002) The skin that we speak: Thoughts on language and culture in the classroom. New York: New Press.
22. Purcell-Gates, V. (2007) Cultural practices of literacy: Case studies of language, literacy, social practice, and power. Mahwah, NJ: Erlbaum.
23. NGSS Lead States. (2013) Next generation science standards: For states, by states. Washington, DC: The National Academies Press.

24. Schoerning, E., & Hand, B. (2012) Language formality, learning environments and student achievement. In: The future of learning: Proceedings of the 10th international conference of the learning sciences (ICLS 2012) (Vol. 2, pp. 154–156). Sydney, NSW, Australia: ISLS.
25. Tobin, K., & McRobbie, C. J. (1996) Cultural myths as constraints to the enacted science curriculum. Science Education 80: 223–241.
26. Moje, E. B., Collazo, T., Carrillo, R., & Marx, R. W. (2001) "Maestro, what is 'quality'?": Language, literacy, and discourse in project-based science. Journal of Research in Science Teaching 38(4): 469–498.
27. Crawford, B. A., Zembal-Saul, C., Munford, D., & Friedrichsen, P. (2005) Confronting prospective teachers' ideas of evolution and scientific inquiry using technology and inquiry-based tasks. Journal of Research in Science Teaching 42(6): 613–637.
28. Rosenau, J. (2015) Evolution, the environment, and religion. http://ncse.com/blog/2015/05/evolution-environment-religion-0016359. Accessed 5/13/2017.
29. Akkus, R., Gunel, M., & Hand, B. (2007) Comparing an inquiry-based approach known as the Science Writing Heuristic to traditional science teaching practices: Are there differences? International Journal of Science Education 29(14): 1745–1765.
30. Moje, E. B., Collazo, T., Carrillo, R., & Marx, R. W. (2001) "Maestro, what is 'quality'?": Language, literacy, and discourse in project-based science. Journal of Research in Science Teaching 38(4): 469–498.

CHAPTER 2

The Culture of Classroom Science: Discourse, Dialog, and Language Practices

When examining classroom science in America, it has become clear that the NGSS are in the process of becoming a defining cultural motif. Forty states are considering adopting the NGSS, and nineteen have done so as on November 2017. The NGSS require that students pose questions, design activities to generate data, and construct claims based on evidence.[1] Accordingly, one of the critical features of the NGSS is argumentative reasoning. This means that evaluating and understanding approaches to science teaching and learning that emphasize the acquisition of this form of reasoning are of growing national importance. While there have been a number of approaches proposed to address argumentative reasoning, Cavagnetto[2] highlights immersion approaches as being of critical importance. In these approaches, efforts are made to teach science as science is practiced, meaning that students must engage in the practices of science, such as the construction and critique of arguments that support explanations of how the world works.

Immersion approaches are the lens through which we will examine the changing culture of the American science classroom. The Science Writing Heuristic (SWH) approach is an example of the immersion approach to argument-based inquiry, and it is an approach which is closely aligned to the NGSS.[3,4,5] Under this approach, students develop questions, design experiments, gather data, and generate evidence to support claims that address their initial questions. This immersion approach, characteristically of the genre, does not utilize set or fixed daily curricula or step-by-step

© The Author(s) 2018   13
E. Schoerning, *Science Culture, Language, and Education in America*, https://doi.org/10.1057/978-1-349-95813-9_2

laboratory experiments. Instead, students and teachers follow templates that guide them in scientific investigation.

When beginning a new unit, the teacher introduces a topic, such as the water cycle. Students are encouraged to ask questions about this topic in an initial brainstorming session. The teacher records these questions, and together the teacher and students choose several questions to explore in class. Students work to further define the questions they generated, and then begin to design experiments that will allow them to answer these questions. Teachers provide assistance in this process in the form of questioning, encouraging dialog, and providing guidance and access to resources. Under an immersion approach, teachers rarely provide direct, factual answers to student questions.

The goal of immersion practitioners is to develop a classroom environment wherein students can acquire and develop the skills necessary to answer their own questions. Once students design their own experiments they perform the experiments and gather data. Teams of students work to analyze these data and develop claims and evidence. Teachers explicitly instruct students in how to construct these scientific arguments. The students' experiments should address a question and allow them to support a claim based on evidence. Students construct these "question–claim–evidence" arguments in their teams, and then present their findings and arguments to the class. Different teams ask questions of each other as they work to construct and critique their arguments and understandings related to the research topic at hand.

The SWH approach is just one immersion approach to inquiry science teaching and learning, but is particularly useful for examining the language and culture of the American science classroom due to previous work establishing its utility in facilitating student–student dialog and argumentation.[6,7,8,9,10,11] Additionally, the SWH approach has been evaluated in the context of a statewide randomized control trial, an unusually high standard of evaluation for an immersion approach, which examined the effects of the treatment on elementary school students in grades 3–5. Results show that students exposed to this type of immersion-based science instruction experienced significantly greater rates of growth on state standardized tests in subjects including math, reading, and science, with greater rates of growth also seen on tests of critical thinking skills. Exposure to this approach was found to narrow achievement gaps between male and female students. The treatment had particularly strong positive effects on underserved demographic groups, such as students receiving free and reduced lunch and black and Hispanic students.[12,13,14]

This immersion approach heavily utilizes many forms of language, both in terms of student writing and as related to the essentially dialogic nature of its classroom environment. However, as Moje, Collazo, Carrillo, and Marx[15] highlight, dialog and student–student talk are rare in traditional science classrooms. Because of the importance of dialog in the immersion classroom, there are particular requirements for these learning environments; in contrast to traditional science classrooms, students in fully NGSS-aligned classrooms must have access to agency and power to openly and fully participate in a truly dialogic classroom.

Here, we examine and describe how students' and teachers' use of spoken language changes as classrooms implement immersion-based science instruction, with the goals of beginning to unpack how these classrooms afford students opportunities for academic growth and success, and of understanding how student agency and power may shift as students are afforded opportunities for dialogic expression.

An NGSS-aligned approach to argument-based inquiry focuses on both enabling negotiation and developing an understanding that scientific argumentation is a particular form of negotiation framed by the discipline of science.[16] Negotiation as seen here is defined as the skillful overcoming of obstacles,[17] and encompasses a complex set of linguistic tools. Accordingly, the use of the term negotiation has significant potential value for the field of science education, including as it does a variety of ways of communicating beyond but related to scientific argumentation by enfolding the concepts of dialog and problem-solving in a larger sense. However, the meaning of negotiation in the context of the science classroom is underexplored.

NEGOTIATION

Useful work related to the concept of negotiation has been accomplished in the field of business, where various aspects of negotiation have been explored, with four main aspects being defined.[18] In negotiation, one diagnoses problems, structures deals, fosters the participation of stakeholders, and builds consensus.[19,20,21,22] These aspects have clear parallels to the science classroom standards and practices described in the NGSS, where it is desired that students should learn to identify problems, find solutions, and build consensus through the discipline-specific norms of scientific argumentation. These three aspects of negotiation have been somewhat addressed in the science education literature through the lens of language and argumentation.

Argumentation has previously been described as the means of scientific discourse,[23] and thus argumentation can be considered a language of science.[24] Research has found that argument enhances students' understanding of scientific concepts, that it improves their understanding of the science process, and that it encourages the development of critical thinking skills by making their thinking processes more transparent.[25,26,27,28] Mastering the language of argumentation is considered a key aspect of developing scientific literacy.[29]

The fourth aspect of negotiation, fostering the participation of stakeholders, while less explored, has particular resonance when we apply negotiation to the language of the science classroom. A goal of science education is to transform students into stakeholders; considerable research in the field has focused on increasing student motivation, and successful approaches to inquiry teaching and learning tend to stress the degree to which students become actively involved in science learning and the science process.[30,31] Research in these areas seems compatible with, but not analogous to, work in this fourth aspect of negotiation, in good part because current research does not address the degree to which negotiation is a language-based process.

Immersion approaches to science instruction focus on enabling all four aspects of negotiation. Researchers have done considerable work to understand how the argumentative aspects of negotiation impact students in NGSS-aligned classrooms through both written and spoken language.[32,33,34,35] This study examined features of negotiation related to the fourth aspect—fostering participation and creating stakeholders. We recognize that power and access change once negotiation is enabled in a classroom. However, we have not fully unpacked the concept of negotiation as it relates to access and power, and thus do not fully understand the mechanisms by which negotiation might afford opportunities for student access and achievement.

STUDENT ACCESS

If students are to have their ideas respected in the classroom, they have a need for access and equity. Once they possess these, their ability to express ideas in the classroom is an expression of power. In traditional science classrooms, students do not possess power, while teachers do so unambiguously. Power here is seen both as authority, in that the teacher has the ability to direct the actions of others in the classroom, and as ability, in that the teacher maintains, distributes, and evaluates knowledge.[36] Power as described here is not strictly dichotomous, being necessarily lost by one group as it is gained by another, but as something that arises and circulates within populations in the Foucauldian sense, as described by Lewis, Enciso, and Moje.[37] In other words, when we say that students have power in NGSS-aligned classrooms, this does not exclude or diminish the power of their teachers; our theoretical framework does not assume a fixed quantity of power in any given social space.

In classrooms utilizing immersion-based practices, while it is recognized that teachers do have an ultimate responsibility and power over their classrooms, students are also empowered. Students engage in power both in terms of authority, as they negotiate what activities should be explored in class and the methods by which they will be explored, and in terms of ability. Students generate knowledge through their investigations, share this knowledge with their peers and their teacher, and work to evaluate both the knowledge they generate and the knowledge generated by others. As a part of this process, they engage in spontaneous scientific talk, which has been found to be rare in most science learning environments.[38,39] The generation of knowledge through argumentative reasoning based on public debates and critiques of claims and evidence is an important way in which students can express power in an NGSS-aligned classroom. In this way, science knowledge becomes understood as something that is negotiated rather than something that is immutable, a truism that carries into an accurate understanding of real-world science and the associated practices of science.

Students' access to power is deeply tied to their access to agency. Agency as discussed here refers to projected agency: the ability of a person to internally look forward toward new paths and directions.[40] In terms of the classroom, agency can be seen as the internal processes of students as they relate to driving the flow of instruction.[41] In a traditional classroom, students have minimal agency,[42] as most choices, such as what to study and in what ways topics will be explored, are made for students by teachers. Low-agency classrooms of this type correlate negatively with academic achievement.[43,44] Additionally, without internal access to agency, students are unable to create meaningful outward expressions of power.

Access to power and agency is a particularly important issue when we consider the success of historically underserved student populations, which have been seen to experience particularly strong positive effects under the SWH approach in the context of the large-scale randomized control trial.[45] Many of the difficulties faced by nonmainstream students are centered in language. The language of science is often seen by these students as exclusive and exclusionary.[46,47,48] Science is seen as something that belongs to the mainstream, not to these students' communities.[49,50,51] These issues related to culture and science have been identified as significant demotivating factors in a wide variety of subgroups in American culture.[52,53,54,55]

One of the reasons mainstream students tend to be more successful in science may be that the language practices they use as home discourse already have much in common with those language practices used in schools. As seen in research on cross-cultural literacy, middle-class white children come to school already possessing many of the skills needed to do well in school.[56] The home language practices of middle-class white people have been shown to involve more of the elements found in scientific argumentation than the home discourse of many other cultural groups. From an early age, children from this group are asked open-ended questions by their parents; parents provide reasoning for why young children may or may not do certain things, and children are expected to provide reasoning for their actions and are asked to provide reasoning for why they may think in certain ways and hold certain opinions.[57,58] These behaviors are not culturally universal. In many cultures, children are expected to learn social norms through observation rather than explicit instruction; children are not asked open-ended questions, are told what to do through direct statements rather than being reasoned with, and are less likely to be engaged in guided or directed Aristotelian-style dialog with their parents.[59] These differences do not indicate that parents from these cultural groups are uninvolved with or uninvested in their children. These cultural groups have different and valid ways of interacting with their children. Various cultural groups within the United States interact with their children differently and utilize different forms of literacy, but these differences when studied by ethnographers do not indicate deficit.[60,61,62,63]

If power and agency are connected to language, and if immersion approaches improve students' access to the discipline-specific language of science, this may help to explain the extent to which students from underserved demographic groups experience success under these approaches.

LINGUISTIC MARKERS OF ACCESS AND POWER

If we recognize that students in NGSS-aligned classrooms have access to agency and power, the question we must ask is as follows: Through what avenues does this access arise? Our current understanding is that the language practices utilized by immersion-implementing teachers and their students as they engage in negotiation provide avenues for agency and power, thus allowing students to develop as stakeholders. Characterizing talk in SWH classrooms as an example of immersion-based classrooms has helped us to understand these avenues, and thereby develop our understanding of negotiation, most particularly that aspect of negotiation that hinges on stakeholder participation. Talk and language are non-neutral; they always convey messages beyond those contained in their vocabulary.[64] Ways of talking include tone, tempo, conventions for dialog exchange, nonverbal signaling, and many other markers.[65]

Many of these nongrammatical features of language serve to encode power. Ways of talking such as tone, tempo, and vocabulary choices indicate both how a speaker chooses to express their own power and their perception of the relative power of their listeners; the use of formal versus informal speech is a way in which individuals can express power.

In examining the differences between immersive and traditional classrooms, there is evidence to suspect that linguistic formality may underlie many of the differences we see.[66] As one's perception of oneself as a stakeholder is directly tied to the degree to which one perceives oneself to have power, we tend to find that speech characteristics related to formality change as negotiation becomes an increasingly predominate paradigm within a classroom. Accordingly, the use of formal and informal speech is different between NGSS-aligned and traditional classrooms, which are often deficient in dialog and negotiation.

Many of the ways of talking Gee describes follow similar formality patterns across languages and cultures. Speech that is low in tone and slow in tempo, for example, is a consistent marker of linguistic formality, whereas certain types of nonverbal signaling, such as laughter, are cross-cultural markers of linguistic informality.[67,68] The presence of conventional turn-taking as opposed to fluid conversation has also been found to be consistently tied to linguistic formality.[69] If the presence of dialog is related to the degree of formality present in spoken language, and scientific argumentation can also be meaningfully framed as dialog,[70] we can understand many aspects of scientific argumentation in terms of linguistic formality.

This allows us to look at argumentation with a lens of access and power that allows complex connections to the larger concept of negotiation. All four aspects of negotiation (defining problems, finding solutions, encouraging participation, and reaching consensus) require dialog. Dialog is not simply the presence of two or more voices in a discussion, but includes complex activities such as reaching consensus, which requires many conversational turns and active engagement in sophisticated linguistic forms such as construction and critique.[71] Accordingly, dialog can be thought of as an occasion where these voices flow through one another, developing, turning, and reflecting upon ideas.[72] Dialog is a form of deeply engaging with the other, allowing for exploration of and complexity in ideas through participation and interrelationship.[73] For dialog to occur, all individuals must have equitable access and power in the classroom, where they feel an ability to express their ideas, and have reasonable expectations that their ideas will be considered and respected. Accordingly, as we seek language characteristics that signal the presence of negotiation within a classroom, there is a need to look for signs of dialog. As the presence of dialog markers increases, language formality decreases. Known markers of dialog include the frequency of dialog interchange between students and teachers and among students, the space permitted for student–student talk, and the presence of speech summarization.

The frequency of dialog interchange, or how often speakers change in the classroom, is a logical indicator of quality dialog when we define dialog as involving flow (as above in Isaacs). Building upon this idea, examining the frequency of dialog interchange both between students and teachers and among the student population may be a fruitful way of analyzing whether negotiation in particular classrooms exists purely between students and teachers or if it disseminates into students' peer interactions. An environment ripe for dialog is somewhat unpredictable; it is negotiated as both students and teachers utilize and develop the statements of students.[74] An important indicator of classroom dialog is that, while the teacher's voice remains significant, it becomes only one of many voices.[75] As teacher voice contributes less to the classroom, more conversational turns and thus more meaningful dialog can occur. When teachers utilize an immersion approach, students engage in more talk.[76] These studies all indicate that student–student talk is a meaningful component of dialog. Speech summarization also indicates the presence of dialog, as in dialog it is necessary that listeners conceptualize the spoken thoughts of others so that their own thoughts can be developed within a culturally appropriate context.[77] Though Vygotsky would argue that this conceptualization takes place internally, instances of observed speech summarization can provide evidence that these internal processes are occurring in both students and teachers.

POWER CONVENTIONS OF LANGUAGE

A goal of science instruction is to give students the tools to engage in science as science is practiced, which includes developing the linguistic skills necessary to participate in the scientifically literate community. Some of these linguistic skills involve aspects related to power, such as the mastery of discipline-specific vocabulary.

One might anticipate that teachers in traditional classrooms would utilize more language characteristics related to power conventions, such as the use of discipline-specific vocabulary, honorifics, and speech with artificially low tones and slow tempos. The banking theory of education would lead one to think that if a teacher frequently utilized discipline-specific vocabulary, this vocabulary would be retained and utilized by his or her students.[78] Similarly, one might think that students might mirror other power-related conventions of scientific language after observing them in their teachers. Finding that teachers who utilize fewer power-related linguistic practices produce students who engage in more of these desired, empowered language behaviors should lead us to question why and how language is shaping the environments of these types of classrooms. When the teacher no longer extensively utilizes power-related language behaviors in the classroom, the degree to which the teacher is the most important person in that classroom is diminished. An environment in which students can develop a sense of power and agency opens up space which was once fully occupied by the instructor. The space is defined by the language behaviors of immersion-implementing teachers, which include the absence of traditionally authoritarian vocalization and body language, the ability to speak freely, and the recognition of the content of speech regardless of its vocabulary. These teacher language behaviors create pathways by which students can gain entry to the classroom dialog.

Students are able to utilize these linguistic pathways as avenues for agency and power. In the absence of teacher speech bearing authority markers such as low tone and slow tempo, or in the presence of relaxed, nonauthoritarian body language, students may feel less concerned about being punished or corrected by authority figures. This gives them greater

opportunity for agency; they may feel they have more freedom to participate in class or think divergently. Similarly, a teacher who values the
content of student speech above the vocabulary used in student speech, or
a teacher who allows students to speak without formal permission, creates
a similar avenue for agency.

These language changes seem to be persistent and self-enforcing.
Although immersion-practicing teachers are instrumental in creating the
environment of an NGSS-aligned classroom, student speech and teacher
speech do not have a strictly dichotomous power relationship. Once an
immersion-practicing classroom environment is established, student
speech also influences teacher speech. If students begin to access the
opportunity provided through a teacher's language behavior by, for example, engaging in on-topic dialog with each other and their teacher during
science class, the teacher's language behaviors will be reinforced. The
teacher will be more likely to speak freely and informally during class if this
behavior results in productive, engaged students. This opens further
avenues for student agency and power, allowing students to become
increasingly involved in and engaged with the conversations that take
place in the science classroom, which further enforces a reduced use of
power conventions in teacher speech, and so on. Similar cycles of positive
reinforcing behaviors have been noted in classrooms in regard to students'
engagement and teachers' instructional behaviors.[79,80]

The styles of teacher speech that characterize SWH classrooms create
an environment that allows the distinct student speech of SWH classrooms
to develop and thrive. When we consider this in the context of negotiation
theory, this is a clear example of how the language behaviors of science
teachers can encourage one of the key aspects of negotiation: fostering the
participation of stakeholders.

POWER LEADS TO DIALOG

As demonstrated by Schoerning et al. in 2015, language behaviors
related to the presence of dialog do not change until after significant
changes occur in the frequency of power-related language behaviors.
This may indicate that in order for the argumentative aspects of negotiation to occur, including defining problems, finding solutions, and reaching consensus, it is first necessary to establish the participation of
stakeholders by inverting the use of power conventions in the classroom.
In order for students to productively and confidently participate in

dialogic conversations, they must have opportunities for both access and power. For teachers to successfully implement NGSS-aligned approaches to science instruction, they must utilize argumentative negotiation in their classrooms, which requires changing student behavior. In a traditional classroom, student voice has a limited role. Students are generally expected to remain quiet unless they are called on by their teachers. When called on, their speech role is generally limited to giving "right answers" or asking "good questions," while other student speech is often seen as disruptive. Teachers may interpret student–student dialog in the context of teacher-led discussion as interruption, even if the students are engaged in discussing a relevant topic. Except in expressly permitted contexts, student–student dialog outside of a teacher-led discussion is often also seen as inappropriate regardless of the dialog's content. Dialog interchange in traditional classrooms tends to be infrequent and purely teacher-directed. Teachers cannot engage in argumentation by themselves; it is necessary that their students participate in argumentation with them. When student speech is an essential part of the classroom, the traditional teaching behaviors described above quickly prove counterproductive. To encourage student speech, teachers summarize student speech, thus demonstrating both to the speaker and their peers that students' speech contributions are important. Teachers engage in more frequent and more fluid dialog interchange, give students more opportunities to speak, and direct students to speak with each other even in teacher-led contexts. Students absorb these lessons and implement these behaviors in discussions with their peers, where significant increases are seen in the frequency of the same linguistic moves used by their teachers.

The direction of these changes is significant. While those behaviors related to power conventions, which change first in the SWH classroom, occur inversely in students and teachers, with the frequency of student behavior increasing as teacher behavior decreases, the pattern of change related to dialogic language behaviors is different. When we consider dialog-related language behaviors, we find that students are learning from teachers as models. Student behavior frequency only increases after teacher behavior frequency increases.

Students do not appear to need to be explicitly taught how to engage in expressions of agency and power. These are skills they presumably bring to the classroom from the linguistic toolsets of their larger lives. However, students do seem to need to be explicitly taught how to engage in dialogic forms of negotiation. While it has been previously supposed that language informality gives traditionally disadvantaged students both the comfort and the confidence that allows them to meaningfully engage in science education,[81,82] this study allows us to see some pathways by which the effect may be occurring. Students are able to gain access and power in the classroom setting through the use of power-related language behaviors that apply to their everyday lives. This may serve to reduce their perceptions of science as an exclusive field, thus increasing participation by increasing student access to agency. Then, through explicitly teaching the language skills involved in building and defending formal arguments as their implementation of the approach improves, SWH teachers give their students the skills needed to engage in dialogic negotiation.

CONCLUSION

There are major changes in the way students and teachers talk in NGSS-aligned and traditional elementary science classrooms; these changes are emergent, and there are patterns by which these changes occur. By actively inverting some of the conventions of formal language, teachers practicing immersion can create avenues for access and power for their students, engaging them as stakeholders in the science classroom, within the first year of practicing the technique. In the second year of technique application, inverting formal language conventions was seen to increase the abilities of students to engage in argumentative aspects of negotiation.

The classroom talk in which students and teachers engage is rich in clues regarding how and why approaches to science teaching and learning can foster or discourage student success. By examining language as a defining aspect of the classroom environment, we will be better able to understand how to successfully implement the NGSS in the science classroom. The NGSS seek to have students learn science as science is practiced, and the way teachers speak in the classroom is related to how students implement science language practices. The way we talk about ideas and practices is often deeply related to how we think of them; the way we talk about science practice can expose our concepts around the NOS.

In the next chapter, we will discuss NOS misconceptions in the science classroom and in the general public as related to science communication, and tie these issues to cultural conflict around science.

NOTES

1. NGSS Lead States. (2013) Next generation science standards: For states, by states. Washington, DC: The National Academies Press.
2. Cavagnetto, A. R. (2010) Argument to foster scientific literacy: A review of argument interventions in K-12 science contexts. Review of Educational Research 80(3): 336–371.
3. Choi, A., Notebaert, A., Diaz, J., & Hand, B. (2010) Examining arguments generated by year 5, 7, and 10 students in science classrooms. Research in Science Education 40(2): 149–169.
4. Hand, B., Norton-Meier, L., Staker, J., & Bintz, J. (2009) Negotiating science: The critical role of argument in student inquiry, grades 5–10. Portsmouth, NH: Heinemann.
5. Norton-Meier, L., Hand, B., Hockenberry, L., & Wise, K. (2008) Questions, claims, and evidence: The important place of argument in children's science writing. Portsmouth, NH: Heinemann.
6. Akkus, R., Gunel, M., & Hand, B. (2007) Comparing an inquiry-based approach known as the Science Writing Heuristic to traditional science teaching practices: Are there differences? International Journal of Science Education 29(14): 1745–1765.
7. Choi, A., Notebaert, A., Diaz, J., & Hand, B. (2010) Examining arguments generated by year 5, 7, and 10 students in science classrooms. Research in Science Education 40(2): 149–169.
8. Hand, B., Norton-Meier, L., Staker, J., & Bintz, J. (2009) Negotiating science: The critical role of argument in student inquiry, grades 5–10. Portsmouth, NH: Heinemann.
9. Norton-Meier, L., Hand, B., Hockenberry, L., & Wise, K. (2008) Questions, claims, and evidence: The important place of argument in children's science writing. Portsmouth, NH: Heinemann.
10. Schoerning, E., & Hand, B. (2012) Language formality, learning environments and student achievement. In: The future of learning: Proceedings of the 10th international conference of the learning sciences (ICLS 2012) (Vol. 2: 154–156). Sydney, NSW, Australia: ISLS.
11. Schoerning, E., & Hand, B. (2013) The discourse of argumentation. Mevlana International Journal of Education 2(3): 43–54.
12. Hand, B., Therrien, W., & Shelley, M. (2014) Argument-based inquiry effects on science, mathematics and reading comprehension: A randomized control study. Learning and Instruction. (under review).
13. Hand, B., Therrien, W., Shelley, M., & Laugerman, M. (2014) Effectiveness of an inquiry-based approach on grade 5 students' critical thinking skills. Elementary School Journal. (under review).

26 E. SCHOERNING

14. Laugerman, M., Fostvedt, L., Shelley, M., Baenziger, J., Gonwa-Reeves, C., Hand, B., et al. (2013, March 7–9) Structural equation modeling of knowledge content improvement using inquiry based instruction. Interactive poster presentation at the Spring 2013 Conference of the Society for Research on Educational Effectiveness, Washington, DC.
15. Moje, E. B., Collazo, T., Carrillo, R., & Marx, R. W. (2001) "Maestro, what is 'quality'?": Language, literacy, and discourse in project-based science. Journal of Research in Science Teaching 38(4): 469–498.
16. Chen, Y.-C., Hand, B., & McDowell, L. (2013) The effects of writing-to-learn activities on elementary students' conceptual understanding: Learning about force and motion through writing to older peers. Science Education 97(5): 745–771.
17. Reznitskaya A., & Gregory, M. (2013) Student thought and classroom language: Examining the mechanisms of change in dialogic teaching. Educational Psychologist 48(2): 114–133.
18. Cobb, S. (2000) Negotiation pedagogy: Learning to learn. Negotiation Journal 16(4): 315–319.
19. Dukes, E. F. (1996) Resolving public conflict: Transforming community and governance. Manchester, England: Manchester University Press.
20. Fisher, R., Ury, W., & Patton, B. (1991) Getting to YES: Negotiating agreement without giving in (2nd ed.). New York: Penguin Books.
21. Raiffa, H. (1982) The art and science of negotiation. Cambridge, MA: Harvard University Press.
22. Susskind, L., McKearnan, S., & Thomas-Larmer, J. (1999) The consensus building handbook: A comprehensive guide to reaching agreement. Thousand Oaks, CA: Sage.
23. Lemke, J. (1990) Talking science: Language, learning and values. Norwood, NJ: Ablex.
24. Tippett, C. (2009) Argumentation: The language of science. Journal of Elementary Science Education 21(1): 17–25. London: Allyn & Bacon.
25. Abi-El-Mona, I., & Abd-El-Khalick, F. (2006) Argumentative discourse in a chemistry high school classroom: An exploratory study. School Science and Mathematics 106(8): 349–361.
26. Crawford, B. A., Zembal-Saul, C., Munford, D., & Friedrichsen, P. (2005) Confronting prospective teachers' ideas of evolution and scientific inquiry using technology and inquiry-based tasks. Journal of Research in Science Teaching 42(6): 613–637.
27. Jiménez-Aleixandre, M. P. (2007) Designing argumentation learning environments. Contemporary Trends and Issues in Science Education 35(2): 91–115.
28. Zembal-Saul, C., Munford, D., Crawford, B., Friedrichsen, P., & Land, S. (2002) Scaffolding preservice science teachers' evidence-based arguments during an investigation of natural selection. Research in Science Education 32(4): 437–463.

29. Cavagnetto, A. R. (2010) Argument to foster scientific literacy: A review of argument interventions in K-12 science contexts. Review of Educational Research 80(3): 336–371.
30. Sevinc, B. (2011) Investigation of primary students' motivation levels towards science learning. Science Education International 22(3): 218–232.
31. Tuan, H.-L. (2005) The development of a questionnaire to measure students' motivation towards science learning. International Journal of Science Education 27(6): 639–654.
32. Akkus, R., Gunel, M., & Hand, B. (2007) Comparing an inquiry-based approach known as the Science Writing Heuristic to traditional science teaching practices: Are there differences? International Journal of Science Education 29(14): 1745–1765.
33. Choi, A., Notebaert, A., Diaz, J., & Hand, B. (2010) Examining arguments generated by year 5, 7, and 10 students in science classrooms. Research in Science Education 40(2): 149–169.
34. Creedy, D., & Hand, B. (1994) The implementation of problem-based learning: Changing pedagogy in nurse education. Journal of Advanced Nursing 20(4): 696–702.
35. Martin, A. M., & Hand, B. (2009) Factors affecting the implementation of argument in the elementary science classroom. A longitudinal case study. Research in Science Education 39: 17–38.
36. Goodman, J. F., Hoadland, J., Pierre-Toussaint, N., Rodriguez, C., & Sanabria, C. (2011) Working the crevices: Granting students authority in authoritarian schools. American Journal of Education 117(3): 375–398.
37. Lewis, C., Enciso, P., & Moje, E. B. (2007) Reframing sociocultural research on literacy: Identity, agency, and power. Mahwah, NJ: Erlbaum.
38. Lemke, J. (1990) Talking science: Language, learning and values. Norwood, NJ: Ablex.
39. Moje, E. B., Collazo, T., Carrillo, R., & Marx, R. W. (2001) "Maestro, what is 'quality'?": Language, literacy, and discourse in project-based science. Journal of Research in Science Teaching 38(4): 469–498.
40. Emirbayer, M., & Mische, A. (1998) What is agency? American Journal of Sociology 103(4): 962–1023.
41. Reeve, J., & Tseng, C. M. (2011) Agency as a fourth aspect of students' engagement during learning activities. Contemporary Educational Psychology 36(4): 257–267.
42. Bandura, A. (2006) Toward a psychology of human agency. Perspectives on Psychological Science 7(2): 164–180.
43. Koenigs, S. S., Fiedler, M. L., & Decharms, R. (1977) Teacher belief, classroom interaction and personal causation. Journal of Applied Social Pyschology 7(2): 95–114.

44. Reeve, J., Deci, E. L., & Ryan, R. M. (2004) Self-determination theory: A dialectical framework for understanding the sociocultural influences on student motivation. In D. McInerney & S. Van Etten (Eds.), Research on sociocultural influences on motivation and learning: Big theories revisited (Vol. 4, pp. 31–59). Greenwich, CT: Information Age.

45. Laugerman, M., Fostvedt, L., Shelley, M., Baenziger, J., Gonwa-Reeves, C., Hand, B., et al. (2013, March 7–9) Structural equation modeling of knowledge content improvement using inquiry based instruction. Interactive poster presentation at the Spring 2013 Conference of the Society for Research on Educational Effectiveness, Washington, DC.

46. Goodman, J. F., Hoadland, J., Pierre-Toussaint, N., Rodriguez, C., & Sanabria, C. (2011) Working the crevices: Granting students authority in authoritarian schools. American Journal of Education 117(3): 375–398.

47. Hodson, D. (1999) Going beyond cultural pluralism: Science education for sociopolitical action. Science Education 83(6): 775–796.

48. Rossato, C. A. (2007) Engaging Paulo Freire's pedagogy of possibility: From blind to transformative optimism. New York: Rowman & Littlefield.

49. Diaz-Rico, L. T., & Weed, K. Z. (2002) The crosscultural, language, and academic development handbook.

50. Duran, B. J. (1998) Language minority students in high school: The role of language in learning biology concepts. Science Education 82(3): 311–341.

51. Tobin, K., & McRobbie, C. J. (1996) Cultural myths as constraints to the enacted science curriculum. Science Education 80: 223–241.

52. Lee, O. (1997) Diversity and equity for Asian American students in science education. Science Education 81(6): 107–122.

53. Lee, O., & Fradd, S. H. (1996) Literacy skills in science learning among linguistically diverse students. Science Education 80(6): 651–671.

54. Rakow, S. J., & Bermudez, A. B. (1993) Science is "Ciencia": Meeting the needs of Hispanic American students. Science Education 77(6): 669–683.

55. Rosenthal, J. W. (1993) Theory and practice: Science for undergraduates of limited English proficiency. Journal of Science Education and Technology 2(2): 435–443.

56. Gee, J. P. (1990) Social linguistics and literacies: Ideology in discourses (2nd ed.). London: Falmer.

57. Choi, I., Nisbett, R. E., & Smith, E. E. (1997) Culture, category salience, and inductive reasoning. Cognition 65(1): 15–32.

58. Peng, K., & Nisbett, R. E. (1999) Culture, dialectics, and reasoning about contradiction. American Psychologist 54(9): 741–754.

59. Neuman, S. B. (1996) Children engaging in storybook reading: The influence of access to print resources, opportunity, and parental interaction. Early Childhood Research Quarterly 11: 495–513.

60. Delpit, L., & Dowdy, J. K. (2002) The skin that we speak: Thoughts on language and culture in the classroom. New York: New Press.
61. Heath, S. B. (1983) Ways with words: Language, life, and work in communities and classrooms. Bath, England: Pittman Press.
62. Purcell-Gates, V. (2007) Cultural practices of literacy: Case studies of language, literacy, social practice, and power. Mahwah, NJ: Erlbaum.
63. Smith, S. S., & Dixon, R. G. (1995) Literacy concepts of low- and middle-class four-year-olds entering preschool. Journal of Educational Research 88(4): 243–253.
64. Fiske, J. (1994) Media matters: Everyday culture and political change. Minneapolis: University of Minnesota Press.
65. Gee, J. P. (1988) Dracula, the vampire Lestat, and TESOL. TESOL Quarterly 22(2): 201–226.
66. Choi, A., Notebaert, A., Diaz, J., & Hand, B. (2010) Examining arguments generated by year 5, 7, and 10 students in science classrooms. Research in Science Education 40(2): 149–169.
67. Gorham, J. (1988) The relationship between verbal teacher immediacy behaviors and student learning. Communication Education 37(1): 40–53.
68. Irvine, J. T. (1985) Status and style in language. Annual Review of Anthropology 14: 557–581.
69. Irvine, J. T. (1988) Ideologies of honorific language. Pragmatics 2(3): 251–262.
70. Duschl, R., Ellenbogan, K., & Erduran, S. (1999) Promoting argumentation in middle school science classrooms: A project SEPIA evaluation. Paper presented at the Annual Meeting of the National Association for Research in Science, Boston, MA.
71. Ford, M., & Forman, E. A. (2006) Redefining disciplinary learning in classroom contexts. Review of Research in Education, 30(1): 1–32.
72. Isaacs, W. H. (1993) Taking flight: Dialogue, collective thinking, and organizational learning. Organizational Dynamics 22(2): 24–39.
73. Schein, E. H. (1993) On dialogue, culture, and organizational learning. Organizational Dynamics 22: 40–51.
74. Boyd, M., & Rubin, D. (2006) How contingent questioning promotes extended student talk: A function of display questions. Journal of Literacy Research 38(2): 141–159.
75. Nystrand, M., Wu, L. L., Zeiser, S., & Long, D. A. (2003) Questions in time: Investigating the structure and dynamics of unfolding classroom discourse. Discourse Processes 35(2): 135–198.
76. Martin, A. M., & Hand, B. (2009) Factors affecting the implementation of argument in the elementary science classroom. A longitudinal case study. Research in Science Education 39: 17–38.

77. Vygotsky, L. S. (1987) Thinking and speech. In R. W. Rieber & A. S. Carton (Eds.), The collected works of L. S. Vygotsky: Problems of general psychology (Vol. 1) (N. Minick, Trans.) (pp. 53–92). New York: Plenum Press. (Original work published 1934).

78. Rossato, C. A. (2007) Engaging Paulo Freire's pedagogy of possibility: From blind to transformative optimism. New York: Rowman & Littlefield.

79. Pelletier, L. G., Seguin-Levesque, C., & Legault, L. (2002) Pressure from above and pressure from below as determinants of teachers' motivation and teaching behaviors. Journal of Educational Psychology 94: 186–196.

80. Skinner, E. A., & Belmont, M. J. (1993) Motivation in the classroom: Reciprocal effects of teacher behavior and student engagement across the school year. Journal of Educational Psychology 85: 571–581.

81. Duran, B. J. (1998) Language minority students in high school: The role of language in learning biology concepts. Science Education 82(3): 311–341.

82. Rakow, S. J., & Bermudez, A. B. (1993) Science is "Ciencia": Meeting the needs of Hispanic American students. Science Education 77(6): 669–683.

CHAPTER 3

Nature of Science Misconceptions: A Source of Cultural Conflict

Common misconceptions regarding the NOS are often reinforced in American classrooms. Traditional classroom science instruction has had little in common with actual scientific practice. Although significant efforts are being made to correct this, with the majority of American states adopting educational policies that actively seek to engage students in authentic scientific practice, implementation levels of these new teaching approaches are mixed. Additionally, the vast majority of Americans were not taught in classrooms utilizing the new approaches, including virtually all adults of voting age. When we review traditional classroom science instruction, we gain a window into how the majority of Americans perceive the NOS.

How Does Traditional Education Produce NOS Misconceptions?

In traditional science classrooms teachers operate as primary sources of authority, with support from formal texts. To be academically successful, students must internalize these sources of authority without internal variation. The process of traditional academic science learning is not one of investigation but of memorization. Much learning takes place in relation to text, both in absorption and replication. When texts themselves are studied, it is notable that, under longitudinal analysis beginning around 1960, science textbooks tend to present an implicit, incomplete, and naïve view of the NOS. Despite moves toward educational reform, no real change has been noted in how textbooks present science.[1]

© The Author(s) 2018
E. Schoerning, *Science Culture, Language, and Education in America*, https://doi.org/10.1057/978-1-349-95813-9_3

As well as exposure through text, students also have laboratory-based science learning experiences, which were traditionally considered advanced or supplemental to text-based classroom instruction. Traditional lab experiences do not involve genuine investigation, but are again exercises in replication. Lab manuals, another authoritative source of text, must be followed precisely in order to generate the correct result to the laboratory "experiment". The primary purpose of these laboratory exercises cannot accurately be said to be experimentation, as they follow carefully designed scripts with set outcomes, but the acquisition of technical skills. The acquisition of these technical skills, such as measuring, is also assessed as successful by the degree to which they are precise, exact, and thus replicable.

In short, traditional science instruction has as its fundamental mode replication, rather than investigation. This is in stark contrast to the NOS, or to the exercise of science as science is practiced by STEM professionals, which is process-based and inquiry-based. The ways in which a student under traditional instruction might internalize a concept of science as a system of facts to be memorized are self-evident. Under traditional instruction, rates of persistence in the sciences were quite low, especially for students who were not white, male, or from higher socioeconomic backgrounds.

When considering issues around education, the NOS, and demographic exclusion, it is interesting to note how research on these issues is and has been conducted. Walls' work in 2012 examines NOS views in young black children. This study is noteworthy due to the extreme paucity of NOS work done in children of color. Walls' review of the literature reveals that, as of 2009, very few studies seeking to learn how students view or conform to what are considered appropriate or accurate views of the NOS included children of color. Examination of study participants reveals that 97 percent of children involved in NOS studies up to the date of review had been white. Most of the small number of children of color who are studied in this branch of the education literature are children who live in the context of predominantly white populations.[2] This means that the literature can provide us virtually no window into NOS beliefs and related practices outside of white-majority communities.

The reality of American life is that, by and large, Americans are significantly segregated by race. This is true economically, geographically within cities, and in our media consumption. Review of the literature reveals that those populations of Americans who do not persist under traditional instruction are by and large almost completely understudied in regard to their views of the NOS, which is only one factor to consider when we hope to examine potential reasons for alienation or acculturation toward the sciences. Interestingly, some of the work that has been done in very young children shows that very young children of color hold fairly strong, positive, process-based views of science, that they see themselves as potential scientists, and that they are engaged in the subject. When do these children reach a point where they can no longer see themselves in the field? What factors cause them to disengage? These are specific and crucial questions for which we only have speculative answers, in good part because we as a research community simply have not collected sufficient data regarding diverse populations. Sufficient efforts have not been made to reach or work in and with nonwhite populations, and questions regarding implicit bias in instruments and data collection are only beginning to be examined.

Perhaps as more people of color navigate the gauntlet of graduate education and professional persistence, and the research community becomes a place of more diverse voices, we will be better able as a group to examine these questions, because we will be better able to voice them. Although the field engages in many attempts at inclusion, not all of them are successful, in part because they do not engage with people of color when developing resources and thus have limited to no feedback to help them realize when the sources being produced are in some way tone-deaf or fail to meet the needs of communities.

For example, a recent project for classroom teachers seeking to bring more diverse perspectives to NOS education recommends role-play exercises for students. All of the role-play exercises but one involve a cast composed solely of white males, excepting a single exercise, in which students are invited to examine the societal treatment of Rachel Carson. In this exercise, they are invited to examine the ways science and society historically discriminated against women, and questioned the veracity and validity of their accounts. The ways in which these trends continue today, in many well-documented systemic biases against women, are ignored. The degree to which students may not be able to identify with or interested in identifying with privileged white males throughout the history of science is not addressed. Additionally and crucially, the simple fact that this white and masculine depiction of history is a choice, rather than an objective reality, is totally ignored.[3]

The history of science includes many people of color, many woman, and, not incidentally, many women of color. There is a great public appetite for these stories, as seen in the runaway success of the recent movie *Hidden Figures*, which examines the lives of three black STEM professionals working in the segregated south, and the historic impact of these women's work.[4]

Men and women of color have been professionally involved in invention and discovery through the history of science, as have white women. But their stories are only beginning to be told in the public educational sphere, and their contributions to the history of scientific discovery and our understanding of the NOS despite the sociocultural pressures they faced are most often viewed in terms of the past wrongs done to their identity groups, rather than current and pressing discrimination. The stories of current children and students of color are issues of theoretical speculation, blame, and pity as often as or more than they are issues of critical, community-facing research.

The evidence for alienation and exclusion under the traditional model is overwhelming. The degree to which the traditional model has produced a society that sees science as a memorizable fixed resource created by authority figures is striking. NOS misconceptions arise naturally when science is not taught as science is practiced, when science is seen as something performed by others rather than a community function, and when science education ignores the contributions of a significant proportion of the community population. While there is a general agreement that the traditional system of science education has led to some less than desirable outcomes, both for individuals and for society, and there have been many explorations of these factors in relationship to the system, there is somewhat less inquiry into the desired products of this educational system and their relationship to the educational mechanism. Considering all the researched and documented challenges implicit in the traditional model of science education, how does the system manage to produce desired products? What are these desired products? Under the traditional model, it is interesting to consider the successful student.

HOW DOES TRADITIONAL EDUCATION PRODUCE SCIENTISTS?

Undeniably, the traditional system of science education has produced many fine scientists. The vast majority of today's working scientists were produced under the traditional system of education. There are virtues to traditional education. A working scientist must be capable of maintaining order and taking notes, so that their work can be replicated. Note-taking and orderly replication are, as previously discussed, key features of traditional education. A working scientist must know a great deal of raw information: what has been attempted and gleaned in their field, and in related fields. The study of external authorities and sources and the ability to retain large amounts of new information are explicit goals of traditional education. The creation of an internal organizational system to allow this information to be recalled and synthesized is a crucial implicit traditional goal.

In the lab, many of the traditional techniques do require intense physical discipline and precision. Without the technical skills gained through traditional laboratory education, where the generation of specific known results allows both students and teachers to assess if laboratory techniques have been performed correctly, it was not traditionally possible for a working scientist to have a reasonable idea if the results of mature, actual experiments could be interpreted as planned, or if they were the result of accidents caused by sloppy technique. Although modern laboratories in many disciplines are increasingly automated, many experiments do still rely on the human touch, and careful and precise actions must be conducted by human hands. This is true across disciplines, and sloppy technique remains a major source of error and concern for working scientists.

Through the traditional model of science education, a working scientist can gain all of these essential skills. However, although discipline, content knowledge, order, and precision are important to professional success in the sciences, when one meets successful working scientists it often becomes clear that these traits do not occupy the heart of their character. The essential traits found in most successful working scientists are curiosity, persistence, and acute sensitivity to the wonder and joy to be found in discovery and exploration. These traits are interesting to consider alongside research demonstrating that many scientists can be diagnosed with positive schizotypy,[5] which I suppose is a way some people might choose to pathologize a set of personality traits. How do these traits develop simultaneously and in balance with an education based in rule-following? Curiosity, sensitivity, and semi-definables such as joy are certainly not skills that are cultivated and developed under traditional models of science education.

Most commonly, it seems, successful working scientists developed these personality traits during their education, or rather, perhaps, preserved them, through a certain creative personalizing of the curriculum. Interestingly, many successful working scientists engaged in varying degrees of curricular rejection or subversion as students. Behaviors such as refusing to do what they considered unnecessary homework, focusing on projects of personal interest at the expense of the curriculum, and other outlier behaviors are not only common cultural tropes about scientists from before 1990, when the exuberant if odd "wacky professor" began to be replaced by the more awkward, withdrawn, and orderly "tech genius", but common enough things for actual scientists to have done in their youth.

During my undergraduate studies, I was told by scientist mentors not to worry overly about my grade point average (GPA), that I was better off spending more time in the lab learning to solve problems than too much time studying for tests. Many of my mentors would describe their own deplorable attendance records, and questionable GPAs, with apparent nostalgia. The majority of working scientists I have met in the generation above mine have shared stories of what might be construed as either delinquency or a tendency toward self-determination, depending on the sympathies of their audience. I find this tendency to be less true in my generation, where at least half of those of us remaining in the field, if not more, are strikingly well-behaved. And now, half a generation later, many budding scientists are coming of age in an academic culture that has placed even further emphasis on rule-following.

Current potential scientists are more likely to be educated by people of my generation, who were more thoroughly winnowed by standardized tests than the generation above mine. Current students have also been more thoroughly subjected to evaluation by standardized testing throughout their educations than students educated ten or fifteen years ago. As access to educational resources has become both more competitive and more essential to students' economic future, the ability of process-driven, problem-solving candidates to progress in the sciences has diminished, unless they are also excellent test-takers.

Some of these tests, most particularly college entrance examinations, have developed extensive industries. Parents in upper-class and upwardly mobile families regularly pay for their children to be coached in how to take these tests. These learning opportunities have been shown to increase standardized test performance. Because standardized test coaching has become such a common part of the college-bound experience for economically privileged youth, it is reasonable to say that this has further increased pressure regarding college admissions for students who are not afforded supplemental coaching. Students like the student I was, who are perhaps able to study from a book rather than receive personalized instruction, are penalized relative to students who can afford more resources, in the form of classes and coaches. Students who cannot afford or access supplementary texts, or students who are not even aware of the cultural norms and practices of the "typical" college-bound student population, are much further disadvantaged.

Standardized testing, which offers a pretense of objective measures of academic preparedness, is, most particularly in the college-bound market, another way that our culture is currently working to reinforce class boundaries. The ways in which the standardized testing industry utilizes and exploits biases is beginning to be explored and addressed, but while there remains a market for standardized test training, it is impossible to say that the socioeconomic bias in the system is seen as a critical issue. Rather, it is clear that this market continues to be seen as a profitable, rather than oppressive, feature.

As an additional concern, due to standardized testing teachers in the K-12 arena find themselves with less space to tolerate unruly if promising students, as teachers' performances are also evaluated by students' standardized test scores. School and district funding are related to student test performance. The degree to which standardized testing has become a high-stakes enterprise for all stakeholders cannot be overstated. Accordingly, rule-following, replicability, and a species of docility have been strongly selected for in the student pool.

What will this mean as we move into the future? Current educational reforms, which do stress a process-based approach to science education, and seek to create learning environments where students can genuinely engage in questioning and discovery, may make a place within the modern educational system for those traits of curiosity and wonder that are so essential to the success and persistence of the working scientist. However, students being educated under this new system are still very young. And even as the education system changes, we are left with a considerable cultural lag period. Current research suggests that, although teachers are developing more sophisticated understandings of NOS, these are not necessarily being reflected in their teaching.[6] However, there are methods being explored that help bring teacher understanding into classroom practice, through explicit instruction and argumentation-based practices.[7] It will take time for classroom teaching practices to shift. Of course, also, for some time to come, most Americans will have been educated under the traditional model. Accordingly, we can expect current public perceptions of science to remain largely the same as they are now, having been crafted by the traditional model. We can anticipate that without widespread intervention, most likely utilizing the medium of informal education, for decades to come science will be seen as a product-based rather than process-based discipline.

SCIENCE AS PRODUCT VERSUS PROCESS: CULTURAL CONFLICT

Even if one accepts that many Americans hold the misconception that science is product- rather than process-based, a natural question arises. What is the harm? Why should this cause societal difficulty or conflict? The answer is profound. Because product-based misconceptions of science create significant epistemic conceptions about science, where science and scientific enterprises become confused with non-process-based ways of knowing. The question of what sources of knowledge are valid and socially acceptable and the question of how we relate to these sources of knowledge are essential to the order and functioning of any society.

In 2017, these questions loomed large in the American civil discourse. Arguments about the legitimacy of knowledge dominated news headlines. Claims that were demonstrably untrue were touted as alternative facts. A substantial subset of the "alternative facts" discourse centered around scientific information on climate change. The scientific consensus on climate change, which is overwhelmingly that the phenomenon is real, serious, and caused by human action, was presented as unsettled. The perspective that "both sides" of the climate change issue should be debated was frequently taken in various news outlets. The United States became the only country to withdraw from the Paris Agreement, a major international negotiation utilizing our best scientific evidence, which seeks to moderate carbon emissions so as to slow the rate of climate change. The goal of the Paris Agreement is to limit global warming to less than two degrees Celsius above pre-industrial levels. This would allow us to delay and hopefully limit the effects of climate change on our planet.[8]

The scientific evidence on climate change is clear. The planet is warming, with serious consequences for most forms of life on Earth. This warming is contributing to species extinctions across the globe, both directly and indirectly, through habitat loss and the increase of severe weather events that imperil ancient migration routes.[9] An increase in global temperatures will result in sea level rise, which will have major impacts on the coastal cities in which most people on Earth live, and will affect the growth and productivity of most major commercial crops, from staple grains like rice, which pollinate in a limited temperature range, to luxury goods like chocolate, which only grows in a narrow band of habitat that is projected to vanish as the climate shifts.[10]

The counterargument presented to this scientific evidence is not a counterargument in the sense one normally assumes, wherein arguments are made within the context of a body of discourse. Although there is a small body of scientific research which does not agree with the consensus of the scientific community on climate change, the arguments presented in the media do not revolve around these scientific conflicts. Rather, the counterargument commonly presented is that scientific evidence is not a legitimate source of knowledge. Scientists are often presented, in this counterargument, as manufacturing evidence for global warming as part of a conspiracy. This conspiracy supposedly seeks to hamstring economic efforts, to reduce the quality of life of the American people, and to secure for the scientists themselves massive grant funding to enable their lavish lifestyles.[11]

It is difficult to understand how any person who had an accurate understanding of how science works, or who was personally acquainted with any scientists and thus the relatively modest lifestyles afforded the vast majority of persons in the scientific profession, could take this counterargument seriously. However, it is undeniable that many Americans have done so, including many Republicans who otherwise score quite well on general tests of scientific knowledge.[12] For many people, a complex web of identity, political issues, economic concerns, NOS misconceptions, and anxiety about changing social conditions have indeed combined to create an atmosphere where the legitimacy of scientific knowledge as a whole is something to be questioned, and perhaps dismissed.

Questions about the legitimacy of various sources of knowledge are louder now than they have been for some time, but they are not new or unique to our age. Societies have always struggled with conflicts between secular and religious sources of knowledge, between different sources of religious knowledge, and between religious and folk sources of knowledge.

Different forms of cultural knowledge can provide competing worldviews and conflict around topics from what constitutes proper childrearing to what constitutes proper conversation.

The cultural conflict that results from these serious NOS misconceptions can be addressed, and it is possible that these conflicts can be reduced. Institutions of formal education, such as the public school system, are working hard to make significant shifts, as can be seen though the increasingly prevalent adoption of the NGSS. These efforts in formal education, however, will reach a demographically limited audience, and are unlikely to reach individuals who have left behind formal education. For most Americans, a high school biology class is the last exposure to science they are likely to have, except through exposure to and engagement with informal educational institutions.

Informal educational institutions, such as museums, libraries, aquariums, and zoos, provide educational experiences to millions of Americans, and have been shown to significantly improve science literacy rates and scientific content knowledge.[13] Is it possible for these institutions to also make changes in process knowledge, that is, in how people understand the NOS? As an added challenge, there are significant American populations that are underserved by current informal education institutions, most particularly those that live outside of the urban areas where most high-quality informal education institutions are found. These populations, perhaps not coincidentally, are often perceived as experiencing among the most significant cultural conflicts around science.

The "othering" of science, and the "othering" of knowledge in general, has emerged in 2017 as a spectrum of "alternative facts", "fake news", and other popular cultural phenomena that deny our shared experience, even our shared perceptions of reality. Reducing cultural conflict around ways of knowing, as described in this subsection, and bringing a more accurate impression of NOS to the American public, is an important goal for social stability and security. If science can be seen as a safe, reasonably reliable, developing process, one in which everyone can engage in in order to develop a shared understanding of our world, and solutions to challenges we experience, it will be easier for people to have objective conversations about the struggles our society faces. There are challenges to be faced in accomplishing this goal, but many can be faced through community outreach.

SCIENCE AS PROCESS: CORRECTING NOS MISCONCEPTIONS THROUGH OUTREACH

Misconceptions are notably difficult to correct, in part because individuals can often hold their understanding of science so deeply and fundamentally that they find it difficult even to imagine that their ideas might be wrong. Informal educational institutions and outreach organizations can play an important role in shifting public perception toward a process-based rather than product-based view of science, because these institutions and organizations have the opportunity to engage people in process-based science outside of formal schooling and as a part of lifelong learning.

Informal education has changed substantially over its history. Museums, an enduring model of informal education, were at one point a space where people could go and see interesting objects related to natural history, anthropology, or archaeology, and hear lectures regarding these subjects. These products were at one point considered the principal products of scientists, and museums, rather than universities, were the primary professional space for working scientists.[14] This is no longer the case. Modern museums are generally staffed by professionals who are extensively trained in education and outreach, and who may be contributors to growing fields of research in informal education and science communication. As a result of their work, museums increasingly provide the public with not only a space to see, but also a space to do. Hands-on exhibits are popular and commonplace. Lectures are increasingly supplemented by higher-engagement forms of contact, such as hands-on workshops and videos with interactive components. The focus of informal education is increasingly on learning through discovery.

The types of authentic scientific engagement that informal educational institutions such as museums are working to incorporate are a way to distribute process-based NOS knowledge into the public domain. Their efforts are, however, geographically limited, as most museums by their nature occupy particular spaces, which tend to be concentrated in populous urban environments. Some museums, such as the University of Iowa Natural History Museum, are addressing this issue by developing mobile components to serve a wider geographic region. The UIowa Mobile Museum, for example, is a retrofitted RV that houses regularly changing interactive touchscreen exhibits, and visits rural communities throughout the state. In this way, tens of thousands of people a year are exposed to supplemental informal education in a community setting.

Mobile museums, which bring a movable "science space" to communities, help to raise the issue of space and science space. Museums and mobile museums provide great resources to people and communities, but there is a self-selecting element around who chooses to enter these types of discipline-designated spaces. There are many people who might enjoy or benefit from learning more about current scientific issues, particularly in a process-based fashion, but who are leery of science learning due to early negative experiences where science learning was seen as either alienating or, perhaps, terribly boring.

Bringing engaging science opportunities to the public through venues outside of discipline-designated space is an interesting growth opportunity. Engaging with the public in public space affords the scientific or science-minded community to work with a potentially broader segment of the population. Science outreach projects, such as science cafes, and The National Center for Science Education's (NCSE's) Science Booster Club program, provide engaging, hands-on science learning opportunities to the general public in fully public venues. In this way, it is possible to bring both science content and science process to individuals who might otherwise self-select away from informal or continuing science education.[15]

Some research has begun to indicate that exposure to informal, process-based science learning can change average community survey responses to NOS items over time and repeated exposure to community events.[16,17] As communities of thinkers interested in science research, science education, and science communication work together to improve and expand public access to science learning, it is possible that we will be able to change public perceptions of science toward a more accurate impression of NOS, and accordingly reduce societal conflict regarding science. With a more accurate understanding of the NOS it is easier for many people to avoid misinformation about science and also easier for many people to view science as a system of discovery rather than a monolithic and fixed worldview.

While a more accurate understanding of what science is and what science does may help alleviate some conflict in society, it should be noted that the majority of Americans already view science generally positively (if, perhaps, less than accurately).[18] Reducing conflict around science itself is a good thing, but there are some areas of scientific inquiry which require particular attention. These are topics around which considerable social conflict has already developed, which have become politicized, and which have become involved in identity-marking by many people. In the next chapter we will explore some of these high-conflict disciplines, the ways discipline-specific misconceptions contribute to the social conflict around these disciplines, and methods for addressing high-conflict misconceptions that may be tied to individuals' identities.

Notes

1. Abd-El-Khalick, F., et al. (2017) A longitudinal analysis of the extent and manner of representations of nature of science in US high school biology and physics textbooks. Journal of Research in Science Teaching 54(1): 82–120.
2. Walls, L. (2012) Third grade African America students' views of the nature of science. Journal of Research in Science Teaching 49(1): 1–37.
3. Radloff, J. (2016) On teaching the nature of science: perspectives and resources. Cultural Studies of Science Education 11: 527–538.
4. Melfi, T. (2016) Hidden Figures.
5. MacPhearson, J., & Kelly, S. (2011) Creativity and positive schizotypy influence the conflict between science and religion. Personality and Individual Differences 50: 446–450.
6. Bartos, S. (2014) Teacher's knowledge structures for nature of science and scientific inquiry: conceptions and classroom practice. Journal of Research in Science Teaching 51(9): 1150–1184.
7. McDonald, C. (2010) The influence of explicit nature of science and argumentation instruction on preservice primary teachers' views of nature of science. Journal of Research in Science Teaching 47(9): 1137–1164.
8. United Nations. (2015) Paris Agreement, English Language Text. http://unfccc.int/files/essential_background/convention/application/pdf/english_paris_agreement.pdf. Accessed online 1/22/2018.
9. Thomas, C., et al. (2004) Extinction risk from climate change. Nature 427: 145–148.
10. AAAS. (2014) What We Know. www.whatweknow.aaas.org. Accessed 1/22/2018.
11. Uscinski, J., Douglas, K., & Lewandowsky, S. (2017) Climate change conspiracy theories. Climate Science. DOI: https://doi.org/10.1093/acrefore/9780190228620.013.328
12. Kahan, D. (2017) 'Ordinary science intelligence': a science-comprehension measure for study of risk and science communication, with notes on evolution and climate change. Journal of Risk Research 20(8): 995–1016. DOI: https://doi.org/10.1080/13669877.2016.1148067
13. Falk, J., & Needham, M. (2011) Measuring the impact of a science center on its community. Journal of Research in Science Teaching 48(1): 1–12.
14. Long, D. (2010) Scientists at play in the field of the Lord. Cultural Studies of Science Education 5: 213–235.

15. Navid, E., & Einsiedel, E. (2012) Synthetic biology in the science café: what have we learned about public engagement? Journal of Science Communication. 11(04) A02.
16. Falk, J., & Needham, M. (2011) Measuring the impact of a science center on its community. Journal of Research in Science Teaching 40(1): 1–12.
17. Norton, M., & Nohara, K. (2009) Science cafes. Cross-cultural adaptation and educational applications. Journal of Science Communication 08(04) A01.
18. National Science Board. (2016) Science and Engineering Indicators 2016. National Center for Science and Engineering Statistics, Arlington, VA.

CHAPTER 4

Culture and Conflict: Science and Social Controversy

While misconceptions about the NOS are related to deep cultural conflicts around science, discipline-specific misconceptions contribute to specific social controversies in interesting ways that are related but not confined to the previously discussed cultural epistemic issues around science as a way of constructing knowledge as opposed to science as a body of knowledge.

To examine discipline-specific misconceptions, we'll next explore the topic of teaching about socially, but not scientifically, controversial topics like climate change and evolution. Traditional methods that ignore the alienating potential of the culture of science are especially unsuccessful when applied to such topics, due to the particular involvement with group and personal identity definition. The ways language and misconceptions impact the teaching of controversial topics will be addressed, and methods by which these topics can be more effectively addressed will be presented.

Evolution Misconceptions

In my work in rural Iowa, I encountered many of the evolution misconceptions defined in the literature, a body extremely well reviewed by Pobiner in 2016.[1] As she notes,

> Cognitively, resistance to evolution beginning in preschoolers is often grounded in three categories of "everyday rules of thumb" that are inconsistent with evolutionary explanations: essentialism (a belief in immutable

© The Author(s) 2018   45
E. Schoerning, *Science Culture, Language, and Education in America*, https://doi.org/10.1057/978-1-349-95813-9_4

categories or kinds), teleology (explanations for the form of something that assumes that its function or design are need-based or invoke an ultimate purpose), and intentionality (assuming that events are purposeful, goal-directed or progressive, and may be caused by an intentional agent).

All three of these underlying misconceptions—essentialism, teleology, and intentionality—are quite common in the general population, and should by no means be thought of as unique to agricultural communities. Indeed, until quite recently, these misconceptions related to evolution were accepted as aspects of reasonable explanations for the evolutionary mechanism by elements of the scientific community.

Consider, for example, Lamarckian evolution. Lamarckian evolution essentially states that if an organism acquires traits during its life in order to adapt to its environment, the organism will pass those traits on to its offspring. Lamarckian evolution was taken fairly seriously as a possible mechanism for how species change over time up until the 1920s, and was not fully discarded until our understanding of genetics advanced sufficiently for the actual mechanisms of biological evolution to be more fully understood. Our understanding of these mechanisms is still a developing process, with research in evolutionary biology currently uncovering new information about basic evolutionary concepts like sexual selection.[2]

The concepts involved in Lamarckian evolution interact substantially with misconceptions around teleology and intentionality. For example, under Lamarckianism, giraffes are seen as having long necks from intentionally stretching their necks, over generations, to reach tree leaves. Here, giraffes are engaging in need-based and intentional evolution.

The human tendency to attribute intentionality and end-directedness to random processes is a well-known psychological trait that expresses itself in many other ways, such as our tendency to see faces or other shapes in patterns or natural phenomena, like cloud or rock formations, and to assume that random happenings are signs, or that they indicate good or bad luck in daily endeavors. These tendencies are strategies that allow us to impose order on the world. They develop in early childhood, in such a cross-cultural fashion that it seems not irrational to speculate that these tendencies have a biological basis.

Young children tend to assume that everything that happens to them happens with intentionality. The question "was it on purpose, or an accident?" occurs frequently in every preschool, home, library, and other venues in which toddlers are able to accidentally or purposefully injure each other, eat forbidden cookies, or break objects. Although young children are usually able to describe when they have done something by accident or on purpose, coming to a place where they can believe others can do things "by accident" takes some children longer to grasp than others. The idea that actions can be unintentional is more developmentally radical than the notion of intentional action.

Similarly, teleological assumptions that forms or designs are need-based occur very early in childhood. As a toddler, my first child frequently informed me that the small pocket in my purse was *for* snacks (a matter of constant concern and debate), when in my opinion it was clearly intended *for* glasses. Really, of course, it was just an inert part of an inert object, onto which we layered conflicting teleologies.

In many ways, the types of questions we ask children in early science experiences can reinforce teleological or intentional assumptions around biology. For example, children are often asked questions about why animals have particular characteristics, such as "Why does a fish have fins?", "Why does a frog have webbed feet?", "Why do whales have blubber?" Correct and expected answers are functional and end-directed. Whales have blubber to stay warm; fish have fins to swim. These are useful answers within particular contexts, but just because they are useful does not mean they are fully accurate, and they clearly lay groundwork for evolution misconceptions. To say that whales have blubber to stay warm is one step away from saying they evolved blubber to keep warm, and that would be a false step. Statements similar to this, such as that lions have manes to protect their necks, are commonly found in many modern science textbooks.[3] When children read these types of statements in textbooks, teleological misconceptions about biology become enforced and supported by a substantial degree of authority.

Similarly, the human belief in essentialism, that categories possess an immutable quality, is also very deeply rooted and applies to many concepts well outside of evolutionary biology. We develop the concept of essentialism as infants.[4] Very small children are often quite interested in sorting activities. Placing things into kinds gives a sense of order to the world, and is often a useful skill that does provide meaningful category separation.

Classification is a skill that is often utilized in early science lessons. For example, very young children might be asked to sort leaves of different colors or characters into kinds, and school-age children might be asked to collect and classify the same leaves in, say, a tree identification exercise. Many early natural history lessons involve placing the natural world into ranks and classes; early lessons in every discipline often rely on binaries. The way that these deeply held and strongly reinforced beliefs and practices around classification could cause people to develop evolution misconceptions should be quite clear. Why would a person not be skeptical of intermediate forms if they had been penalized in school for an inability to classify something as belonging clearly in one category or another? If many of our early lessons involve immutable categories of biological objects, it is of course difficult for many people to move toward an acceptance of biology as process.

In many ways, it is evident that deep psychological concepts formed in early childhood and reinforced in early biology lessons are tied to some of the most common misconceptions people hold around evolution, and there is evidence that these ties are more than theoretical. When I would talk with people in the field about why they had difficulty with evolution, of course some would bring up religious conflict, but it was actually more common for people to tell me that evolution "felt" wrong.

Evolutionary theory failed to correspond to the way these people internally organized and understood the world in profound and unsettling ways. They expressed this discomfort with some clarity, with many self-reflective individuals independently tying this discomfort to categories and ways of thinking they had held from early childhood. They would say things like "I don't know, I always thought of the world that way. You see one kind of animal, and another, from when you are a kid, and they stay that way".

On a couple of occasions people brought up other concerns they had: that typically when a living thing was outside of known categories it was in a state of disease and deformity. People had seen birth defects, or deformed frogs and other animals in nature. One woman noted to me that "when a thing comes out different, it's not that it gets wings, generally. When something comes out different, it tends to come out wrong".

Entertaining notions that were in violation of the order people imposed upon the world caused them emotional and psychological discomfort. Perhaps crucially, I was often asked by people how it would benefit them to better understand evolution. How would it make their lives any better? Nobody wants to go through an uncomfortable mental process for no reason.

Finding ways to reduce this emotional and psychological discomfort is an important element of evolution education, as is presenting the relevance and applicability of evolution to individuals and communities. But before we address some potential solutions and applications, it's worth exploring content misconceptions and their origins around another socially heated scientific topic.

CLIMATE CHANGE MISCONCEPTIONS

Climate change misconceptions are complex and diverse, and have ties to other classes of misconception. Perhaps the most significant of these general misconception groups are those concerning gases. Well-studied climate change misconceptions include critical misunderstandings of the greenhouse effect, which requires understanding that there are materially and chemically different types of gases, beliefs that carbon dioxide gas forms a distinct layer in the atmosphere, which is located beneath the "healthy" atmosphere, and common misconceptions confusing and/or conflating climate change with the gaseous ozone layer.

In this latter case, misconceptions generate specific confusion regarding the seriousness of climate change. In the 1980s, the hole in the ozone layer was a widely publicized issue, which was addressed by the banning of classes of chemicals known as chlorofluorocarbons (CFCs) in an international effort. After these chemicals were banned, the hole in the ozone layer began to repair itself more quickly than anticipated. Many people hold the misconception that this highly publicized issue, and the successful environmental restoration that followed concerted political action, is identical to or congruent with climate change: either that this issue is/was the same as climate change and is now resolved, or that issues around carbon dioxide and other greenhouse gases will be or are as simple to resolve as those chemicals which had harmed the ozone layer.[5,6]

An important element of this confusion, I suspect, is that many people hold a deep and fundamental misconception that all gases are the same. Atmospheric gases do not look different to the naked eye. When we encounter pollution or other unpleasant elements in our atmosphere, we almost always experience them visually. We take notice of clouds of smoke or odd-colored vapors but we have almost no practical or immediate experience with unwholesome or deadly clear gases.

The exception to the rule, at least in the public eye, is carbon monoxide. The public health dangers of exposure to this gas are well known, and most homes contain carbon monoxide detectors. When children are educated about the dangers of carbon monoxide, in many regions of the United States they are taught that carbon monoxide is more likely to be located "low" in a home, such as in or near a basement. It should be noted that this does not paint an accurate picture of how the gas behaves: carbon monoxide rapidly disperses throughout a closed room. Children are taught this information due to the fact that in many regions of the United States malfunctioning home furnaces are the most likely source of carbon monoxide, that home furnaces are mainly located in basements, and that in many regions of the United States basements and lower floors of homes are more often poorly ventilated. These pieces of information dovetail in an interesting way with misconceptions about carbon dioxide and other greenhouse gases forming a separate and distinct layer below the "healthy" atmosphere.

Deep preconceptions about layered gases may also be developed in another area of early childhood education: fire safety. Any person who received an appropriate fire safety education as a child will be able to recall rhyming phrases such as "stay low and go". Children are taught that smoke rises, and that they may be able to crawl in a layer of safe air that lies beneath dangerous, unbreathable air to get out of a building and away from a fire. Diagrams and coloring books show the layer of smoke above the safer, breathable air beneath.

Early childhood fire safety and carbon monoxide education are memorable childhood experiences for many people. They often include classroom visits from firefighters, a class of personage greatly admired by most children. They may involve hands-on contact with firetrucks, a class of vehicles that for most children hold near-mythic significance. And these educational experiences often involve gifts, such as safety-related coloring books, and perhaps toy firefighter hats. These factors combine to make these educational experiences significant for young children, which is a good and desirable outcome. Children need to learn fire safety. However, these experiences may also serve to create a deep, unintentional misconception about the behaviors of gases: that gases are only or mainly present in bands or layers.

This argument should not be taken as a condemnation of early childhood safety education, which is crucial, saves lives, and provides many important implicit lessons, such as the reliability, safety, and trustworthiness of rescue workers. What this argument seeks to do is reveal a potential source of major misconceptions around gases, which come most significantly into play when trying to explain and understand the significant gas phenomena around global warming.

Perhaps if efforts are made to explicate common misconceptions about gas behavior, as well as their potential sources in early childhood, people will be more able to understand and identify their misconceptions around gases. Additionally, the consideration of the sources of these misconceptions may be a highly relevant and largely unexplored factor contributing to some people's defensiveness regarding gas misconceptions. Who among us is entirely comfortable questioning assumptions instilled in us as young children, for our safety, by trusted and friendly authority figures like firefighters?

Gas misconceptions, especially when combined with the notion that environmental problems concerning gases are easily resolved, readily set the stage for important scientific misconceptions about climate change. But how are these misconceptions at all related to those frequently associated with evolution, and do these relationships provide for us any signposts as to how we might reduce or eradicate these misconceptions?

How Are These Classes of Misconceptions the Same?

Misconceptions about both climate change and evolution can arise from ideas learned in childhood from trusted authority figures both inside and outside the classroom. Early childhood education around public safety can contribute to misunderstandings about gases. The types of actions young children are taught to take to fight climate change can be so simplistic and ineffective that, as has been seen in anti-drug programs, it is possible that they create the opposite of their intended effect, and discourage conservation in some adults. Some forms of religious education serve to create misconceptions about evolution, as do some forms of early science learning, in which living things are sorted and separated into kinds, and the separateness of these kinds, rather than their relatedness, is emphasized.

This paints a dreary picture, much more so than intended. None of these authority figures are attempting to deliberately instill misconceptions in young children: not teachers in secular or religious contexts, not public officers, and not parents working to educate their kids. Consider the following point: a topic on which a person has no misconceptions is not likely to be a topic about which a person has been perfectly and thoroughly instructed, but one on which a person is perfectly ignorant.

Misconceptions most often arise naturally, as unintended consequences of learning. The child or adult who holds many deep and unquestioned misconceptions is not likely to be an uninterested child or adult, but one who is curious, and has spent a fair amount of time considering the world around them. Misconceptions are often naturally produced and explored through the processes of learning through argumentation.[7] Many people who are quite well educated hold fairly bizarre ideas without even realizing it, due to conclusions they've incorrectly drawn from observing natural phenomena.

To illustrate this concept, the most reasonable person to embarrass is my own self. I held the quite significant misconception that electricity is a liquid until I was nearly twenty. I was made to realize the shocking truth by some friends who inadvertently uncovered my deep confusion when we were working together on a repair project. I kept referring to wires as if they had the same properties as tubes, and had unusual concerns about spills.

Eradicating my serious misconception about electricity was a not unemotional process. I was willing to believe what my friends told me. They were, after all, budding electrical engineers, so it seemed likely that they knew what they were talking about. My friends were generally trustworthy sources of information, and I did not think they could be simply pretending to be as astonished at my misconception as they appeared to be. However, to understand electricity more accurately, I had to re-examine phenomena I had studied and drawn false conclusions around in childhood.

Curious about electricity, I had gotten in a great deal of trouble when I was very small, perhaps three or four years old, for attempting to smash open some batteries with a hammer. Hammer confiscated, I was told I could not smash batteries because they contained dangerous acid. The answer was clear before me. This liquid, whose further exploration was forbidden, was electricity.

The rest of my childhood proceeded with a stable, and thoroughly incorrect, model of how electricity functioned. Batteries contained liquid volumes of electricity proportionate to their size. Power lines transported electricity along roadsides, much as pipes carried water below ground. Light switches operated on a gate and sluice basis. The extraordinary danger of electrocution in water? The reason I could not play Tetris in the bathtub? A potentially violent reaction between two liquids, such as I had often seen illustrated in television shows about chemistry.

It is evident to any reader who does not share my childhood delusion that my deep misconception about electricity required selective perception. I must have disregarded information I encountered that ran contrary to my mental model. When watching science programming, I must have ignored or possibly internally argued with those segments teaching actual information about electricity.

However, as an adult, when confronted directly about my misconception by other, more knowledgeable adults whose expertise I trusted, my view of the world shifted. This change made me feel deeply uneasy. What else did I not know? What else in the world did I deeply misunderstand? In what other ways did my mind work against itself, to maintain myself in ignorance?

When speaking with other friends, it seems this type of serious childhood misconception is not uncommon. I have known people who believed in forms of spontaneous generation until young adulthood, or that the sun revolved around the Earth, traversing our sky each day as it passed overhead. I have known several people who believed that women carried lots of tiny babies in their bodies, all the babies they might ever have, and that sometimes a baby would wake up and start growing if a woman slept too close to a man, an appealingly complex notion with interesting parallels in medieval scientific thought.

To return to the thread, what is important to note here is that complex content misconceptions are common, are widely held among educated adults, and have emotional content as they often arise in early childhood. Misconceptions are often the result of information incorrectly learned or applied from trusted authority figures, and can be a sign of curiosity and the completely accurate and appropriate application of early stages of the scientific method.[8] This, of course, is exactly why the scientific method calls for continual application and review of new information, and that review is often what is missing around our misconceptions. We do not even see our misconceptions as requiring review. Perhaps because we do not even see them.

When I faced a serious misconception in my worldview, I experienced self-doubt and confusion. However, I was able to get past these negative emotions in part through experiencing a positive emotion. I enjoyed learning new content information about electricity as I did a hands-on project with my friends. I was very fortunate in that I was able to have trusted experts to teach me examples of how electricity actually worked in my moment of crisis, and to provide examples that clearly proved my previous assumptions to be incorrect. In addition to the enjoyment of learning new content knowledge, I was able to come around to enjoying intellectual uncertainty, to try to bring a new level of questioning to my lived experience, with the hope of identifying other possible misconceptions.

While it is dangerous to generalize too much from one person's experience, my experiences with misconceptions have given me tools that have helped many other people encounter information that is contrary to their preconceived notions. When teaching, my goal is to bring a sense of accomplishment to the discovery of a misconception, rather than embarrassment, and to help others move from the destabilizing, world-tilting stage of misconception discovery to the stability of the scientific method. The greatest strength of the scientific method is its ability to embrace change, its ability to come to logical conclusions that may in fact prove hilariously wrong over time, and to revise these conclusions in the face of new information and ideas.

This approach to addressing misconceptions, which acknowledges the emotional processing inherent in misconception work for people of all ages, is particularly important when dealing with socially contentious issues, such as climate change and evolution.

Effectively Addressing Socially Contentious Misconceptions

In the case of climate change and evolution, misconceptions are as likely as any other misconception to be deeply rooted in childhood learning and experience. They are perhaps even more likely to have strong roots in identity, as they address big picture questions around the relationship of humanity to the natural world, the origins of life, and the well-being of the world. Accordingly, these questions have elements that are often addressed in religious as well as secular education. These questions have strong epistemological elements and reflect deep issues in society.[9]

In the potentially embarrassing misconceptions described in the previous section, my friends and I were in a way fortunate. None of our misconceptions were terribly associated with our adult identities. However, imagine the situation of a person for whom the rejection of climate change has numerous identity implications, for example a person who works in the petroleum industry and whose community's economic health is also based on this type of employment. Imagine a person who has spent a good deal of time at the local café arguing against climate change with their friends and peers. There would be substantial identity-based costs to reviewing and possibly revising their misconception.

Working through misconceptions that are relatively untied to personal identity is already an emotionally challenging process. Working through misconceptions where a person must question elements or aspects of their identity is even more difficult. Coming to a resolution with new understandings also requires changes in one's understanding or application of identity elements.

There is a certain branch of popular understanding wherein climate change and evolution are seen as issues that have political sides, people who are more conservative cannot accept the scientific opinion on these issues, and these issues are inherently liberal or contain liberal bias. There is also a certain line of thought that evolution is in conflict with a religious identity, a point which will be explored further in the next chapter.

These polarizing forms of discourse are not helpful for misconception correction. When working with conservative communities on these issues, for example, it is important and useful to provide appropriate sources of trusted authority. There are many prominent conservative, libertarian, and independent politicians, industrialists, and thinkers who accept climate change. It is essential to provide people who are struggling with identity-tied misconceptions with examples and role models who have corrected the misconception, or who have a record of accepting scientific evidence related to climate change, in line with their identity. By doing so, it is possible to mitigate the emotional work that must be done to process, review, and revise the misconception.

Traditional forms of science education do not account for this emotional work. While engagement in the scientific method can be very useful for correcting NOS misconceptions, which we discussed in the previous chapter, for sensitive content misconceptions a simple and straightforward discussion of evidence in favor of climate change or evolution may prove less than useful. If a person is presented with evidence that is contrary to a deeply held misconception which is tied to their identity, that person is more likely to want to protect their identity than they are to dispassionately evaluate the evidence presented. The presentation of the contrary evidence is likely to be read as an emotional attack. Unsurprisingly, the presenter of such evidence is likely to be perceived as an attacker, a person with hostile, rather than helpful, intentions. People become angry and defensive when they are in pain and that pain is not acknowledged. They become avoidant and sensitive to further injury. In other words, they are unlikely to evaluate the evidence placed in front of them, much less utilize it to revise misconceptions they may hold. There is, in fact, a strong possibility that they will actively reject the evidence, seeing it as a personally offensive entity rather than information.

Teaching a person about a socially sensitive topic is in some ways different from teaching a person about a topic like photosynthesis, which is unlikely to be deeply entwined with their identity. The presentation of evidence is important in all education, but so is the manner of presentation. In my work with climate change and evolution, I have found that communities are often very receptive to learning about these socially controversial ideas if they are presented in a localized, personally meaningful context. It is also important to be aware not only of common factual misconceptions, but also of common emotional and identity-related attitudes toward the topic under discussion.

For example, when I began my work in Iowa, I found it somewhat difficult to find institutions or groups that would partner with me to provide public informal evolution education, because it was perceived that this was a topic which would cause community conflict and result in complaints. Thanks to several collaborators at the University of Iowa Natural History Museum, most significantly Tiffany Adrain, I was able to develop an exhibit openly discussing evolution that was successfully shown to tens of thousands of Iowans at sites across the state, with an overwhelmingly positive participant response. The reason I think people responded so positively to this exhibit is that it was completely grounded in local identity issues, whereas evolution is often seen as a concept not only separate from but antagonistic to local identity groups.

Evolution is seen by some people as a worldview that counters their values, that is depersonalizing, and that makes life feel less significant, meaningful, and miraculous. People who adhere strongly to creationism do not do so based solely on quasi-scientific evidence or biblical literalism. They do so at least in part because a created world, with mankind as the center and apex of God's creation, generates very positive feelings about themselves and the world. This is and has been true of many science-minded people throughout history, such as John James Audubon, famed for his lithographs of birds. His religiosity was noted by his family and can readily be seen in his texts.[10] When one reads his writings, an evident theme is how dearly he perceives God's love in the profusion, beauty, and variety of birds. When Audubon saw a bird, he did not simply study its form and line. He experienced a bird, emotionally, as a gift of the Lord. When he admired the natural beauty of living things, there is a sense in the text that he felt personally cared for by a loving God.

Anyone who cannot see the emotional appeal of such a worldview is unlikely to learn anything applicable from this book. Anyone who considers creationists as merely wrongheaded, rather than people who hold beliefs for the same complex reasons we all do, discards any chance to learn how to meaningfully communicate with people who are deeply invested in a worldview that may be foreign to them.

When I have spoken with creationists and with creationist faith leaders, the predominant message I receive from them is not that they believe in a created world based on the work of so-called creation scientists but that their beliefs are based in a sense of majesty, awe, and love of the Lord. To many of them, rejecting scientific evidence and embracing a creationist worldview is an act of faith and worship: a dedication of the natural world to God's creation.

These are people who care about big ideas and big feelings, who care about meaning and purpose. Why would a person want to abandon a worldview like that to one of black and white, cold scientific facts? Why would they want to be reduced to just another animal?

Of course, that is not how people who adhere to the scientific consensus on evolution or the age of our world perceive things, at least not most of them. Many people who love science, myself included, do so because in the scientific worldview we also find big ideas and big emotions, a sense of majesty and awe, a love of the universe. However, when we attempt to teach people about the concepts that hold so much meaning for us, we have a tendency to adhere to the scientific method and present the evidence, rather than the feeling. To reach diverse audiences, we must recognize that these emotional elements of science, our feelings and emotional responses to science, are as essential to our scientific beliefs as the scientific evidence we have encountered and incorporated into our worldviews. It is that experience and worldview entire that is compelling and inclusive, to a degree that the facts alone will never be.

When I taught about evolution in rural Iowa, I wanted the feeling to come through. We assembled a collection of fossils from four periods in Iowa's geologic history. Many of the specimens had been found by farmers on their property. When I saw people begin to withdraw from the exhibit, threatened by the negative associations of an encounter with evolution, I worked to communicate to them that this was an exhibit about our shared history as Iowans. I emphasized the importance of Iowans like them in the discovery of our history, the richness and beauty of our geologic history, and the fact that this evolutionary heritage is a heritage that belonged to them. In other words, I worked to put participants in the emotional center of evolution as a lived and local heritage, rather than on the emotionless exterior of a dispassionate theory.

The response to this approach was positive. I regularly saw people who were quite cautious of the topic begin to engage with the content with enthusiasm. When I stressed this attitude to the many volunteers who worked with me on this project, and helped them to practice and develop this approach to the exhibit, they reported that they saw the same kinds of positive changes in participant reactions. In one case, I had a woman come up to me at the exhibit with the purpose of engaging in confrontation. She told me she didn't believe in evolution, with the body language of a person who looked interested in a confrontation. I smiled at her, and kept my body relaxed.

"This is something that belongs to you", I told her, and invited her to handle the pieces.

She engaged with the fossils and other materials for about ten minutes, and I answered her questions about the pieces, their origins, their ages, and their relationships. I answered her questions about how these ages and relationships were determined. I was friendly and warm, and let the woman lead the experience, where, with less defensive participants, I might be inclined to guide. There was no further confrontation. Later that hour, the woman led a number of children from her church over to the exhibit and allowed them to experience the exhibit, which they enjoyed. She stayed there, watching, to guard them. It was evident she was concerned that learning about evolution would involve emotional harm, and perhaps intended emotional harm, to the children in her care. From her experience alone at the exhibit she had been able to evaluate the scientific content, which she clearly considered appropriate for her church group. It must not have been the factual information that led her to exercise caution in the encounter.

The kind of sensitivity we built into our exhibits has been crucial to the success of NCSE's Science Booster Club program, which grew out of the work I and many others did in Iowa. The realization that the misconceptions around climate change and evolution are not simply fact based, but tied to deeply held ideas that are bound up in personal identity and early childhood experiences, implies that a simply fact-based educational approach is unlikely to be effective.

When addressing misconceptions in these content areas it is essential to do so in the context of local identity. This can be done by providing role models who adhere to local identity norms but embrace the topic of discussion. It can also be done by grounding the discussion in local needs, experiences, and history. Giving people a sense of ownership and significance around these controversial concepts is absolutely essential.

Many educators across the United States are already using approaches like this. There is a growing sense of the importance of place in climate change education programs in particular, as seen through trends in the proposal calls of federal granting institutions.[11,12] A localized approach is less commonly seen in evolution education, where many people and educators continue to experience a great deal of tension around religious worldviews. It is quite possible that place-based evolution education, emphasizing local involvement and local history, will be a helpful alternative route that may help some educators avoid theological debate. However, for those who cannot avoid theological elements in the science classroom, there is the next chapter.

NOTES

1. Pobiner, B. (2016) Accepting, understanding, teaching, and learning (human) evolution: obstacles and opportunities. Yearbook of Physical Anthropology 159: S232–S274.
2. Schurko, A., Neiman, M., & Logsdon, J. (2009) Signs of sex: what we know and how we know it. Trends in Ecology 24(4): 208–217.
3. Keep, S. Personal communication, NCSE staff. 1/10/2018.
4. Gelman, S. (2004) Psychological essentialism in children. Trends in Cognitive Sciences 8(9): 2004, 404–409.
5. Rocklov, J. (2016) Misconceptions of global catastrophe. Nature 532: 317–318.
6. Oversby, J. (2015) Teachers' learning about climate change education. Procedia Social and Behavioral Sciences 167: 23–27.
7. Hamza, K., & Wickman, P. (2007) Describing and analyzing learning in action: an empirical study of the importance of misconceptions in science learning. Science Education. DOI: https://doi.org/10.1002/sce.20233
8. Settlage, J. (2007) Prognosis for science misconceptions research. Journal of Science Teacher Education 18(6): 975–800.
9. Jasanoff, S. (2010) A new climate for society. Theory, Culture, and Society 27(2–3): 233–253.

10. Walsh, J. (1904) Audubon, the naturalist. American Catholic Historical Society 15(1): 8–21.
11. National Academies of Science, Engineering and Medicine Gulf Research Program. (2018) http://www.nas.edu/gulf/grants/education-2018/index.htm?_ga=2.122715915.1897361745.1515019309-1835389935.1513714083. Accessed 1/23/2018.
12. National Oceanic and Atmospheric Administration Grant Impact Page. (2018) http://www.noaa.gov/office-education/elp/impacts. Accessed 1/23/2018.

CHAPTER 5

Science and Religion: Meshing and Conflicting Worldviews

Here we come to an exploration of the conflict broadly hinted at in earlier chapters: that of science and religion. When science is perceived as alienating, and is often misunderstood as a competing worldview of absolutes, with epistemic characteristics congruent with those of religious belief, it is not unexpected that conflict between science and religion should result. This conflict is broadly promoted by the media, but to what degree does it actually reflect the American cultural landscape?

SCIENCE AND RELIGION: STATE OF THE UNION

Although conflict between science and religion is not universally experienced by people of faith or individuals who hold a purely humanist or scientific worldview, some subsets of both groups do exhibit prejudices against each other. Specific survey data presented in this chapter allows us to see which religious groups and belief systems experience conflict with science, as well as the degree of conflict experienced. The reasons behind these conflicts are explored with an eye toward cultural perceptions of science and the perception of the scientific culture toward various faith groups.

The first and most important message to absorb from this chapter is that the perception of conflict between science and religion is actually quite low in most populations of most countries around the world. While this conflict is played up in the media, research has repeatedly demonstrated that it is simply not a major issue in societies worldwide. A recent

© The Author(s) 2018
E. Schoerning, *Science Culture, Language, and Education in America*, https://doi.org/10.1057/978-1-349-95813-9_5

survey (Borneo, 2015) done across teachers with and without biology degrees in thirty countries finds minimal difficulty reconciling science and religion in nations throughout Europe, Asia, and Africa, with only respondents from a small number of countries in the Middle East and Northern Africa (MENA) region reporting concern. Interestingly, it is only these countries, including Tunisia, Algeria, and Lebanon, which report concerns around evolution similar or greater to those of the United States, where approximately half of the surveyed population finds potential conflict between evolution and creationism.[1]

The proportion of Americans who perceive conflict around evolution do not necessarily also perceive a larger or general conflict between science and religion. Baker's 2012 survey work in the United States demonstrates that, although nearly half of Americans do not have an issue with creationism being taught in the science classroom, this acceptance of "equal time" propositions is not necessarily related to religious commitment. Indeed, a variety of sources appear to link the acceptance of an interest in "equal time" solutions to the central American value of fairness, wherein the majority of issues are perceived as having two sides.

A small minority of people surveyed by Baker considered science and religion incompatible. Interestingly, these people fell almost entirely into two groups: biblical literalists and atheists. Two-thirds of atheists reported incompatibility, favoring science. Less than a quarter of biblical literalists reported incompatibility, favoring religion. This indicates that the vast majority of religious Americans do not perceive an impasse between the disciplines, with a higher probability of conflict coming from people who self-identify explicitly as non-religious.

When Baker dug down in his data to look at the perceptions of the group perhaps most often stereotyped as anti-science—poor, uneducated religious persons—he found only a twenty-three percent correlation with perception of an essential conflict between science and religion. Of people who perceived incompatibility of science and religion, favoring religion, nearly seventy percent reported the perception that "scientists are hostile to religion".[2]

These numbers agree with information NCSE has gained from work in the field. We find that, in many communities, people are somewhat hesitant to engage with informal evolution education not because they perceive a religious conflict, but because they are concerned about social conflict. Specifically, they are afraid that they will be treated poorly by members of the scientific or educational community because of their religious beliefs.[3] Their social concerns are not related to their religious community. They are, in general, unconcerned that they will face negative social consequences from learning about evolution and unconcerned about the impact of learning about evolution on their spiritual well-being. As a point of additional interest, we have found that members of religious communities are generally highly interested in obtaining diverse educational experiences for their children. It is absolutely critical to note that the primary social concern we encounter in the field is related to concerns about participants and/or their children being mocked, judged, or degraded directly, or through the mocking, judgment, or degradation of their affiliated groups in educational materials around evolution.

This concern is valid. A review of popular educational videos around evolution finds that surprisingly many of them include, at the very least, "soft knocks" on the religious. While it is unusual to find extended anti-religious diatribes, the inclusion of small amounts of material that could be logically perceived as anti-religious is common. Materials featuring Neil DeGrasse Tyson on PBS point out that religion is unnecessary, and he has been quoted saying, "…God is an ever-receding pocket of scientific information that is getting smaller and smaller as time moves on".[4] The Dawkins foundation produces a wide variety of evolution education materials, but these also often include anti-religious messages, which is unsurprising when one considers some of the things Richard Dawkins has written, such as "The God of the Old Testament is arguably the most unpleasant character in all fiction: jealous and proud of it; a petty, unjust, unforgiving control-freak; a vindictive, bloodthirsty ethnic cleanser; a misogynistic, homophobic, racist, infanticidal, genocidal, filicidal, pestilential, megalomaniacal, sadomasochistic, capriciously malevolent bully".[5]

There is no necessity for evolution education materials to feature anti-religious messaging. When reviewing such materials, I have often been struck by the educational pointlessness of these inclusions. They do not build on any of the learning objectives currently defined by state standards for science education and they do not contribute to an understanding of evolutionary theory. While many of these "soft knocks" are one-liners in otherwise entirely educationally directed materials, they do contribute to the creation of a culture of conflict around evolution. They certainly validate the not uncommon perception that scientists are hostile to religion, a perception that is backed up by survey data of scientists. While the majority of scientists are accepting of faith, there is a significant minority that is more hostile to the religious than surveyed religious groups are to scientists.[6] There is some qualitative ingroup recognition of this problem,[7] which is complicated by associations made by the New Atheist movement, which explicitly attempts to pair atheism and science.[8]

When members of religious groups develop the rightful and logical perception that educational materials around evolution are often subtly but explicitly anti-religious, they are often put into challenging situations regarding their own learning. Work by NCSE in the field to dialog with families who embrace creationism has revealed that the majority of these families would like greater access to information about evolution. Parents report wanting their children to be able to access evolution education for several reasons. They care about the quality of their children's education in general and want them to have well-rounded educations. They are aware that STEM professions offer good employment prospects. Additionally and significantly, many parents report that they know an understanding of evolution is an important cultural marker. They want their children to be able to move smoothly through the world, under-standing information about evolution even if they don't agree with it, so that they have the social skills they need for economic upward mobility. They are aware that not all Americans or citizens of the world share their beliefs, and want to be educated about what they may perceive as other belief systems.

However, if the educational materials available actively challenge their belief system, their feelings of exclusion and desire to protect themselves and their children from emotional harm do cause families and individuals to avoid these educational materials. As perceived by members of these communities, avoiding education does cause communities to become more insular and to have less access to economic opportunity. The tension between the desire to obtain education and the desire to avoid harm causes people to make difficult decisions.

Causing people to avoid education, of course, is in direct opposition to the goals of the many undoubtedly well-intended people producing evo-lution education materials that contain "soft knocks". These people and institutions want to provide resources to underserved groups and to increase educational opportunities for all children. Many of these individ-uals might, perhaps, express dislike of insular religious communities that avoid education, and hope that their materials would contribute to these communities' integration with larger society. Unfortunately, the inclusion of "soft knocks" is in direct conflict with this goal.

When looking at survey data for another socially controversial scientific topic, climate change, we see surprisingly similar trends. The majority of Americans want more information on this topic. Relatively small groups are polarized on either side of the issue, with the majority of people falling into a broad middle, which involves generally low emotional investment. Loud, minority voices in the community work to describe scientists as untrustworthy sources of information who do not have the best interests of Americans at heart. And many sources of climate change education contain information that can be perceived as contrary to community values or economic interests. Research in science communication has helped us to understand ways in which the scientific community unintentionally offends or frightens many demographic groups in the way we present climate change research.[9,10]

Returning to the general theme of conflict between science and religion, research leads us to two important points: generally speaking conflict is quite low, and when conflict is present the source and perpetuation of conflict more often originates in scientific rather religious groups. The assumption of conflict limits our understanding of both science and religion, as it forces us into a particular epistemological frame.[11] Much of the current conflict model is based in an ethnocentric model of religion, which exclusively utilizes particular strains of Christianity as the sole basis for understanding all religious frameworks and experiences.[12] Exploring ways that science and religion exist without conflict can broaden our understanding of these topics and their relationship.

SCIENCE AND RELIGION: ROADS TO RECONCILIATION?

Why is it that so many religious groups and individuals do not experience conflict around science? Reviewing the views of the major American religious denominations provides an interesting perspective. The Clergy Letter Project provides us an excellent window into these worldviews. The Clergy Letter Project, led by Michael Zimmerman, is a way for individual religious leaders as well as religious denominations to express support for evolution.[13] Denominations that have signed on include the United Methodist Church (worldwide), and the Presbyterian Church (United States). The letter has been signed by over 14,000 Christian clergy, over 500 Unitarian Universalist clergy, 500 rabbis representing many branches of the Jewish tradition, and dozens of Buddhist clergy. In conversation with Zimmerman in January of 2017, I was able to inquire why Muslim clergy were not involved in the project. He informed me that this was more a matter of access than anything else, that he did not have access to Muslim faith leaders through his social or interfaith networks. Similarly, he reported that this was why the number of Buddhist signatories was fairly modest. While Islamic thought of course presents a diversity of opinions, the majority of Muslim faith leaders in the United States are embracing and accepting of science, as well as informed of historical and present contributions by Muslims to the scientific community.[14]

As demonstrated by the sheer number of signatories to the Clergy Letter Project, there are a significant number of religious denominations and institutions that do not experience cultural conflict with science. And of course it is worth noting that the faiths mentioned in relation to the Clergy Letter Project represent only a small part of the diversity of world religions. The tendency in the United States to talk about the conflict between science and religion is misleading, and assumes a very narrow interpretation of religion: Abrahamic religion with a particular school of biblical interpretation. One way of interpreting biblical text is to take the stance that one is not interpreting the text, that taking the plain literal interpretation of the text requires no interpretation. Of course, there is no such thing as reading text without interpretation; every person reads text through the lens of personal experience, including their time, place, and the assumptions and values of their culture. To take the stance that there is one true, literal, and singular interpretation of the Bible is a position that both requires and denies intense interpretation. This style of biblical interpretation is not uncommon in some Christian denominations in the United States, and is often called biblical literalism.

If one is engaged in biblical literalism, it is difficult to square the biblical account of creation with current scientific theories on the origins of our planet and our universe. Although this is perhaps the best-known conflict, there are many other stories in Genesis that present significant challenges to the scientific consensus, such as Jacob's methods of livestock breeding. The split-reed method described in Genesis 30:37-30:43[15] has, to my knowledge, never been presented as a serious "alternative genetics" argument, although this would be a logically consistent argument with those made by creation scientists.

Biblical literalism does create fixed bounds for its adherents, which is doubtlessly part of the psychological appeal of such a belief system, which offers clear, if not easy, answers to most questions. In the absence of biblical literalism, there are many ways to perceive and construct truth and ways of knowing.[16] The ways in which various religions do this, and fit science into that worldview, differ. In the discussion below I will only touch on the Abrahamic religions, and hope that, like Zimmer's Clergy Letter Project, I will be excused for the lack of breadth in my focus. There are many non-Abrahamic religions which I am neglecting, entirely due to my own lack of experience and education on their relationships to the scientific worldview. I justify my conscious omission by noting that, in America, I am unaware of any non-Abrahamic religious group being reported as engaged in public conflict related to science or science education.

Interestingly, not all science-accepting worldviews adopt the culturally dominant separate magisteria pathway, which is frequently cited by people trying to reconcile science and religion from the perspective of scientists. Among many adherents of the Abrahamic religions, it is common both historically and presently to see scientific knowledge as a subset of religious knowledge, inspired by religious knowledge, or otherwise not only compatible, but intimately related to religious knowledge.[17] This has been documented from a variety of religious Christian, Jewish, and Muslim perspectives.

From Christian thinkers, it is useful to examine this quote from Cardinal Newman, written in 1868, back when society was first wrestling with Darwin's theory:

> I do not fear the theory so much as he seems to do—and it seems to me that he is hard upon Darwin sometimes, which [sic] he might have interpreted him kindly. It does not seem to me to follow that creation is denied because the Creator, millions of years ago, gave laws to matter. He first created matter and then he created laws for it—laws which should *construct* it into its present wonderful beauty, and accurate adjustment and harmony of parts *gradually*.[18]

Here, Cardinal Newman is writing to a fellow clergyman, with an intellectual response that was not at all uncommon across Christian denominations. Christianity, as the dominant faith of the Western world for well over a thousand years, has a rich and complex relationship with the development of Western science that is well beyond the scope of this chapter. But for every story reminiscent of the suppression of Galileo, it is equally possible to find an instance where Christian clergy have celebrated, upheld, or directly contributed to scientific progress. What is unusual, in the Christian intellectual tradition, is to see science and the study of the natural world as a pursuit intrinsically separate from the rest of intellectual life.

The Jewish tradition has viewed the sciences as deeply related to Torah learning throughout its history, with authorities embracing the study of nature and the natural sciences as appropriate and beneficial throughout Jewish history.[19] The positions taken by major thinkers on this point are quite strong, with Leibowitz, writing in the twentieth century, stating, "From the point of view held by the Jewish religious person... the confrontation between science and religion does not exist".[20]

Many religious Muslims see deep connections between the Koran and modern scientific knowledge. Verses from the Koran are interpreted as describing the structure and development of the human embryo, complex astronomical information, and other scientific findings that exceed the technological capacities of human society at the time of the revelation. The historical contributions of the Muslim world to science and mathematics are well known, and many modern Muslim communities deeply value education and make significant contributions to world knowledge. Again, the general tendency within the religious framework is not to see science as a separate way of interpreting the world but as a way of knowing deeply connected to, and in this faith tradition to some degree stemming from, sacred text.[21,22]

Of course, there have been times, places, and particular branches of all three of these faiths that have been, sometimes violently, opposed to science and secular learning of all kinds. I readily acknowledge that the past few paragraphs present a positive face of the interaction between science and the Abrahamic faiths. However, I think that is a face worth presenting, particularly because it so consistently makes arguments that are not at all aligned with the way the scientific community often views the relationship between science and the Abrahamic religions.

How can we utilize these insights for effective science teaching? Clearly, they suggest that there is not one right way to incorporate a process-based understanding of science into a coherent worldview. Many people feel a sense of fairness around the idea of "teaching both sides" of an issue, and throughout American history our society has broadly interpreted science and religion as two sets of answers for the same phenomena.[23] However, this dualistic approach diminishes the richness of the human experience, and renders invisible non-binary solutions. In classrooms where cultural conflicts have arisen around science, it is possible that teaching diversity may help to provide a solution. When a variety of pathways to resolution are made available, the potential for conflict resolution substantially increases. There is also evidence that preservice education on NOS issues can help student teachers feel more confident teaching topics like evolution, as they are able to better understand and appreciate the diversity of ways of knowing.[24]

This type of understanding, where we are able to empathize with other people's experiences even though we do not agree with their positions or conclusions, is invaluable to broadening participation in science, both in the classroom and in the community. In the following and final portion of this text I will explore my experience with this concept and its execution in the field. Although the first half of this text involves a degree of intellectual and philosophical complexity and underlies the approaches I used in the field when working in Iowa, the actual implementation of these ideas involved a wildly different skill set, primarily social and logistical. These challenges underlie every kind of fieldwork, and every kind of fieldwork is performed by people whose personal situations, relationships, and backgrounds inform their choices. None of us are actually objective observers. I find I am unable to describe the fieldwork without describing myself, as it was a part of me.

Notes

1. Clement, P. (2015) Creationism, science, and religion: a survey of teacher's conceptions in 30 countries. Procedia, Social and Behavioral Sciences 167: 279–287.
2. Baker, J. (2012) Public perceptions of incompatibility between "science and religion". Public Understanding of Science 21(3): 340–353.
3. Bolger, D., & Ecklund E. (2017) Whose authority? Perceptions of science education in Black and Latino churches. Rev Relig Res. DOI: https://doi.org/10.1007/s13644-017-0313-6
4. The Science Network. (2011) The Moon, the Tides and why Neil DeGrasse Tyson is Colbert's God A Conversation about Communicating Science. Accessible online: http://thesciencenetwork.org/programs/the-science-studio/neil-degrasse-tyson-2. Accessed 1/3/2018.
5. Dawkins, R. (2006) The God Delusion. Bantam Books.
6. Baker, J. (2012) Public perceptions of incompatibility between "science and religion". Public Understanding of Science 21(3): 340–353.
7. Long, D. (2010) Scientists at play in the field of the Lord. Cultural Studies of Science Education 5: 213–235.

8. Cimino, R., & Smith, C. (2011) The New Atheism and the formation of the imagined Secularist community. Journal of Media and Religion 10: 24–38.
9. Kahan, D., et al. (2016) Culturally antagonistic memes and the Zika virus: an experimental test. Journal of Risk Research 20: 1–40.
10. Leiserowitz, A., Maibach, E., Roser-Renouf, C., Rosenthal, S., & Cutler, M. (2017) Climate change in the American mind: November 2016. Yale University and George Mason University. New Haven, CT: Yale Program on Climate Change Communication.
11. Evans, J., & Evans, M. (2008) Religion and science: beyond the epistemological conflict narrative. Annual Review of Sociology 34: 87–105.
12. Gironi, F. (2010) Turning a critical eye on Science and Religion: theological assumptions and soteriological rhetoric. Method and Theory in the Study of Religion 22: 37–67.
13. Zimmerman, G. (2018) The Clergy Letter Project. http://www.theclergyletterproject.org/. Accessed 1/4/2018.
14. Personal communication, Arlington, VA, January 2017.
15. Leiber, D. (1999) Etz Hayim, Torah and Commentary. The Rabbinical Assembly, New York, NY.
16. Hathcoat, J., & Habashi, J. (2013) Ontological forms of religious meaning and the conflict between science and religion. Cultural Studies of Science Education 8: 367–388.
17. Reiss, M. (2010) Science and religion: implications for science educators. Cultural Studies of Science Education 5: 91–101.
18. The Letters and Diaries of John Henry Newman, edited by C.S. Dessain and T. Gornall, vol. XXIV (Oxford: Clarendon Press, 1973), pp. 77–78.
19. Sacks, J. (2011) The Great Partnership: Science, Religion, and the Search for Meaning. Schocken Books, New York.
20. Neusner, J., & Avery-Peck, A. J., (2001). The Blackwell reader in Judaism. In ch. 17, Orthodox Judaism, Brown. 255. Oxford, UK.
21. Al-Azm, S. J. (2007) Islam and the science-religion debates in modern times. European Review 15(3): 283–229.
22. Edis, T. (2009) Modern science and conservative Islam: an uneasy relationship. Science and Education 18(8): 885–903.
23. Preston, J., & Epley, N. (2008) Science and God: an automatic opposition between ultimate explanations. Journal of Experimental Social Psychology 45: 238–241.
24. Stears, M. (2012) Exploring biology education students' responses to a course in evolution at a South African university: implications for their roles as future teachers. Journal of Biological Education 46(1): 12–19.

CHAPTER 6

A Case Study in Transforming Communities: The Science Booster Club Program

Introduction to the Case Study

This chapter is not written in a traditional academic fashion because I don't know who on Earth actually reads traditional academic prose for anything approaching pleasure. Pleasure in and of itself, I mean. For those who read a good deal of academic prose, I think there is a certain good feeling that arises from encountering the discourse: the pleasurable feeling of ingroup recognition.

> Ah!
> Says the brain
> (In that I mean one can indeed feel chemicals being released. The body relaxes, the pupils dilate)
> The sign of my people!

I would argue, however, that this sort of ingroup recognition, which of course implies outgroup exclusion, is only incidentally related to the pleasures to be found in quality and worth. Learning to see these things (quality and worth) as removed from our complex ingroup preferences is a challenge indeed. That process is what lies at the heart of this story.

And this story is really quite interesting. It is the story of how, in less than two years, a program was developed from the ground up that is energizing and mobilizing communities all across America and serving to educate tens of thousands of people about what had been seen as highly divisive topics: climate change and evolution.

© The Author(s) 2018   71
E. Schoerning, *Science Culture, Language, and Education in America*, https://doi.org/10.1057/978-1-349-95813-9_6

In order to tell you this story the form must to some degree meet the function, and the form of this story was one of intense human connection. Accordingly we begin, enjoying a roughly quarterly breakdown, as we progress along a narrative arc.

FEBRUARY 2015: THE SETTING

February of 2015 was an evil time in Iowa. Temperatures stayed under twenty below for weeks. The ground was covered with an inch of ice, rotted where it wasn't slick. Utterly impossible to play outside. Nearly impossible to go outside, particularly with an infant and a toddler. Most particularly impossible with a sick infant, leaking foul fluids from every orifice, miserable and wailing. I spent my time this month reading Leviticus, attempting to maintain something resembling a human posture despite the constant pain in my gut, and, whenever the children were sleeping, frantically working or applying for work.

I needed work quite badly. My husband had been removed from the home at the end of January, but he was under no court order compelling him to support us. I had part-time work, I had contract work, but by no stretch of the imagination did I have mortgage-paying work. And, as my husband had told me, I was entirely unemployable—overeducated, uninteresting, stupid, and lazy besides.

Alas! I feared this might be true, but I did have to keep trying. If I could find work, if I could find a way to support my children and keep us in our home, the odds were much greater that I would be able to keep my children. I had the most terrible fear that I would lose the children, for I had been told repeatedly by my husband that he would take them away from me if I did find work, which would almost certainly be out of state. He said anyone would see it this way, for while I might be, objectively, a physically adequate mother, it was clear to anyone who knew me that I was a corrupting moral influence.

I struggled and the ice melted. The many applications I wrote out in February began to bear fruit in early March. I received many requests for interviews.

It is quite true that I am perniciously overeducated. When I was fourteen years old I left my family home in the greater Chicago area to attend the Illinois Mathematics and Science Academy, a state boarding school. I then completed a bachelor's degree in biology at Illinois Wesleyan University, and immediately afterward began doctoral work in microbiology.

I had achieved my dream and been admitted to Tulane's infectious disease program, and was happily ensconced in the back top corner of a moldering mansion in New Orleans' Garden District. I had so wanted to live in New Orleans! And I did, for a time, until the hurricane. For a while after the hurricane, too, but I found it impossible to progress in my studies, and so I followed my lab to Arizona State University. There, after passing my comprehensive exams, I left the bench to focus on the human experience of science. On my twenty-sixth birthday I defended my doctoral thesis, which focused on how vocabulary use affects science learning.

Quite significantly, was my finding. By focusing on concept acquisition and eliminating discipline-specific vocabulary, I significantly changed my students' course outcomes.[1] Students who were non-native English speakers, black students, and Hispanic students no longer failed my course at disproportionate rates. When tracked into upper-level courses, students who had experienced my low-vocabulary treatment outperformed their traditionally educated peers.[2]

I was able to further develop this interesting line of research at the University of Iowa, where I was granted a postdoctoral research fellowship. Over the course of three years, I identified many aspects of language that significantly impact student learning in the sciences. And I was able to demonstrate that as teacher use of these linguistic aspects changed, student speech also shifted, sometimes in important, non-intuitive ways.[3] The results of my work were published in a respected journal.[4]

In my search for employment it turned out that there were people who found this work less insignificant than my husband had. In particular, there were many people who struggled with reaching different populations— who wanted, for example, for their institutions to fail less than eighty percent of their Hispanic students, or to improve their freshman retention. And then there was one group that was trying to reach everybody.

It wasn't an academic job. I was more than a little intimidated by the glamor of the outfit, which I had certainly heard of before. And the people. Their names and faces, I knew them as well, and as sources of authority. Always quoted in all sorts of publications.

They had me out to interview. My good suit, which had been purchased for me when I was twenty years old and applying to graduate schools, fit quite well even after two children due to the slimming combination of terror and poverty. I know how to put on a good face. I still owned a little good jewelry.

The decision? Almost immediate. She would have me. They would have me!

I would start in April. This was mid-March.

MARCH–APRIL 2015: INTRODUCTION TO THE PROBLEM

I had a good deal of work to do before I started work. For one thing, I needed to obtain regular childcare. Although I would be working from home, the companionship of my very small children in virtually every instance was an impediment to productivity. Finding a regular daycare would be expensive and difficult, especially considering the ages of my children and my masses of legal bills. I also needed a reliable phone. My ex had removed me as an account holder on my own phone line, and there was nothing I could do about it. I also knew that he used my current device to monitor my contacts and probable location. There was, additionally, the threat that he might turn it off at any time.

With the promise of regular income both the problems of childcare and communications seemed surmountable. I could certainly afford a phone now, but it quickly became clear to me that I could not afford to place a one-year-old and a three-year-old in a licensed daycare facility. In my region the cost for this was approximately eighteen hundred dollars a month, more than the average woman's monthly after-tax income. With shame and disgust at the degree to which my life had come to resemble a Lifetime movie, I clicked on a Craigslist ad for a home daycare, which promised availability in the necessary age ranges. The next day we visited, and met the kindly, well-intentioned older woman who would watch my children, along with an absolutely unreasonable and almost certainly unsafe number of other children, for one thousand dollars a month.

My ability to work obtained, if in a less than satisfactory fashion, it was time for me to address another problem central to my employment. As the new Director of Research and Social Organizing for NCSE, I had the mission to develop outreach strategies that would reduce covert threats to science education.

At this point, it is certainly worth saying a bit about NCSE, as it will continue to feature quite prominently in this narrative, which is not entirely about my survival but also, and perhaps more interestingly, about my work.

The roots of NCSE began to develop in the late 1970s and early 1980s, when bills promoting "scientific creationism" came to prominence in several states. Concerned citizens organized into groups that came to be known as the "Committees for Correspondence", an endeavor eventually involving all fifty states that was coordinated by Stanley Weinberg. In 1981 members of several of the Committees for Correspondence founded NCSE, which was legally incorporated in 1983.[5]

Since then, NCSE has been the nation's only nonprofit dedicated to defending science in America's public schools. The organization has been instrumental in major legal battles to keep creationism out of public schools, such as the Kitzmiller trial. We provide advice and assistance to teachers, parents, and students who encounter creationism or are pressured not to teach evolution in their science classrooms. In 2012 climate science was added to our mission, as we began fielding increasing numbers of complaints from teachers under fire for teaching about global warming. Generally speaking, if you run into a problem with the teaching of scientifically sound yet socially controversial topics, NCSE is the organization that will have your back.

I was thrilled to be associated with the organization, whose work I had admired for some time. I came to know of them and to make use of their resources during my time in Iowa working with rural teachers, who often struggled with the exact issues NCSE served. Through their experiences I learned that the importance of having a mechanism in place to deal with open anti-science conflict simply cannot be overstated. There are many communities where this remains a serious threat to the quality of public education.

However, NCSE's research had shown that these flare-ups represented an alarmingly small proportion of total anti-science activity in the American school system. Surveys had found that thirteen percent of American teachers actively teach creationism in the science classroom, although it is not legal to do so in most states. These numbers alone demonstrate the need for NCSE's continuing mission but don't tell the whole story. An astonishing sixty percent of teachers largely avoided any discussion of the socially controversial topic of evolution, hedging the topic, avoiding or minimizing instructional time on evolution, or failing to teach the subject to a level that meets state standards.[6] This left only a little more than a quarter of American teachers providing accurate and sufficient evolution instruction to their students. Additionally, we found that only half of teachers were accurately teaching climate science, our other key issue.[7]

We had tools for fighting overt threats. But we had little if any means for dealing with the subtle social pressures that were causing the majority of American teachers to avoid teaching science. NCSE was investing heavily in new programs, targeting both teachers and the general public. My colleague Minda Berbeco was developing a new network to provide teachers with support and resources to teach the essential scientific topics included in our mission. I would focus on community organizing. My history of successful outreach to groups traditionally underrepresented in the sciences was seen as having strong potential for general public outreach. My belief in access to science education as a tool for social and economic justice was seen as compatible with the larger mission and mindset of NCSE.

If I was to begin dealing with these covert threats, I needed to understand what I was getting into. What was the state of science education in Iowa? I knew something about this issue from the experiences I had gained during my postdoctoral research.

As part of my language work, I visited classrooms all over the state, spending time with dozens of elementary school science teachers. The research group to which I had been attached, led by Brian Hand of the University of Iowa, was looking at how classrooms changed as teachers adopted inquiry-based approaches to teaching. I ran professional development workshops for the teachers and visited their classrooms regularly to see how they were doing, to score instruments regarding their implementation of inquiry-based methods, and also to collect data on how their linguistic behaviors and those of their students were or were not changing.

It was very interesting being embedded in these classrooms over months and years. I collected a tremendous amount of data, but I believe that I was able to do this in a minimally invasive way. Much of the time I was in classrooms I tried to be of help to teachers. I worked to provide support, positive feedback, and manual labor, rather than to openly prioritize their classrooms as a lab space for data collection. I tried hard to show my respect to the teachers. They were all professionals whose experiences and skills were different from mine. Their experiences and skills gave them a crucial perspective on their classrooms that I wanted and needed to hear. I wanted to hear what they thought was important. I wanted to hear their stories. And as I became accepted in different schools, as teachers began to believe that I was indeed a researcher who respected them as professionals, the stories they told me tended to change considerably.

After I spent six to nine months in a community, the teachers there would often become much more honest and open with me. Teachers with whom I had spent considerable time working on inquiry-based lesson plans would literally close the book; they would set aside their binders and gently inform me that they did not actually teach that kind of material in their classrooms. The kind of material they did not teach was, more often than not, anything even peripherally related to evolution.

This was my first time personally encountering anything like this, either in terms of information censoring or science denial. As a child, I was raised in an atheist home and fed a steady diet of Asimov. Although I cannot distinctly recall spending much time on evolution in a science class until I was an undergraduate, I can't believe that I gained all of my understanding of the theory from science fiction. It must have been part of the backdrop of my earlier education. From a young age I thought of evolution, a concept absolutely fundamental to a modern understanding of biology, life history, and the natural world, as a theory whose beauty and applicability were equaled only by its everyday, commonplace acceptability.

It shocked me that teachers were actively avoiding the topic of evolution, even though state standards called for its inclusion in the curriculum. But you never get any more good-quality information out of people if you react to their revelations with shock, disgust, or dismay. They clam right up afterward. The first time a teacher shared these elements of her teaching practice with me, I think I was able to keep my cool pretty well. It did quite shock me that first time.

I worked to draw teachers out when they revealed these types of stories. The reasons they were avoiding evolution turned out to be surprisingly consistent. In most but not all cases it was unrelated to their personal religious beliefs. Rather, this type of censorship was driven by concerns about community backlash, even if none was currently present, and a lack of support to deal with community pressure. Teachers were concerned by even the possibility of conflict in their communities; there didn't need to be a history of conflict in the community around evolution for teachers to be leery of teaching on the subject. Stories in the national media highlighting conflict in other communities appeared to have as strong a suppressive effect as actual conflict in a local community. I heard frequently from teachers that they did not believe their administrators would protect them if community conflict did occur.

This climate of avoidance and fear did not appear to be mollifying anti-science elements in rural Iowan communities. The knowledge that school administrators and teachers would bend the science curriculum to suit them appeared to encourage these groups to brand new topics as morally unacceptable. In fact, the story that first led me to learn about NCSE and their resources was not in regard to evolution at all but a related, and perhaps even more serious, scientific issue: deep time.

Deep time, or as some people call it, extended geological time, or geological time, or perhaps, simply, time, is an essential concept if one is to understand modern scientific theories on any number of topics. Evolution doesn't make sense unless it takes place against a backdrop with simply oodles of time. The urgency of modern climate science findings does not seem sensible without an eye toward an extended picture of the past. The physical structure of the solar system, let alone the physical structure of the universe, all of those vast distances, are impossible to comprehend without an appropriately vast timescale.

One of the younger teachers I worked with told me one day, very upset, that she had been counseled by her administration not to teach a lesson plan about erosion to her third graders. There were concerns that talking about erosion would lead to conversations about how the Earth's landscape changed over time, which was really the problem. There were concerns that it was inappropriate to teach the children about geologic time. There might be community backlash if that type of thing were discussed in school, due to an influential local pastor preaching young Earth creationism.

If you have not encountered the theology of young Earth creationists, it is worth noting some of the basics of their beliefs. People who ascribe to young Earth creationism can belong to a variety of Christian denominations but they are united by a literal interpretation of the Bible, most particularly focused on a literal interpretation of the creation stories in Genesis as an expression of God's will and majesty. Young Earth creationists believe our planet to be only around six thousand years old. They believe the entirety of the Earth's history to be recorded in the Bible, and that the age of the Earth can be deduced through that text's genealogical information.

Many people who believe in a created world, rather than one formed by natural processes, still accept the scientific scale of time. They will argue that, when the Bible speaks of the days of creation, these days can be understood as an allegory, or that the days of the Lord are not like our days. Many if not most people of faith who love the Bible do not interpret the text literally, but strict literal interpretation of the text is an inarguable element of faith for some. The pastor in my area who was preaching young Earth creationism was one of this group, and made a point to say not only that he was a young Earth creationist, but that he was a "short day" young Earth creationist who explicitly believed that each day of creation described in the Bible took place over a 24-hour period.

The educators I worked with were intelligent, caring professionals. They had a variety of religious beliefs, and many were people of deep faith. But I have no doubts on this score: none of them had a moral issue with teaching scientifically accurate information about geologic time. The teachers I worked with, even those who openly described themselves as creationists, were acutely embarrassed to be under pressure to censor their curricula to avoid reference to geologic time. They were aware that this was a level of science denial that was simply not acceptable in larger American society. They were entirely conscious that this type of censorship would cause their communities to be viewed as pathetic, primitive, and ignorant by the outside world. They also knew that to deny children in their communities access to this information was to cripple their educations.

Which is why teachers and administrators did not want these stories getting out. It was clearly very inappropriate, very risky, to tell these stories to outsiders. Particularly egg-headed university types like me. People do not want to be viewed as ignorant. They do not want to be viewed with disgust. But they don't want to deal with conflict in their communities either.

It is exactly this type of situation, where shame and secrecy run against community cohesion, that permits all sorts of nasty things to grow. In this instance the dichotomy between insider and outsider and the desire to keep the peace was allowing covert threats to science education to flourish well beyond the bounds of even what one might consider typical religious concerns.

The stories teachers told me about science denial in their communities, related to both the pressures they experienced and their practices, were immensely important and influential to me. I spent a lot of time thinking about them. Teachers began to tell me stories of science denial, generally speaking, in tandem with another kind of story. Both tended to emerge after about six to nine months of regular contact. Once trust was established I began to hear about problems related to money.

I learned, for example, of how very few schools intended to implement any long-term changes to the curriculum as a result of the research trial in which we were currently engaged. The teachers told me that they participated only for the money and would find some other study to enroll in afterward for whatever supplement that research offered. I found that several of our schools were actually simultaneously enrolled in competing trials, a confounding factor of which I am sure relatively few researchers on either team were aware. It was all done for the money. The schools and the teachers were in desperate need of money.

This was shocking to me. Growing up in Illinois in the late 1980s and early 1990s, Iowa was always presented as the absolute pinnacle of education in the United States. Even the tests I had to take, that ended up determining so much of my future, were the Iowa Tests. The Iowa Test of Basic Skills was widely utilized over the course of my education. Almost every student in American public schools took these tests. It was clear to all of us children that Iowa was a major player in education. I always figured it must be very nice there. I remember how happy I was when I moved to Iowa, how happy it made me to think that my children would go to Iowa schools.

My belief in the pleasantness of Iowa and the quality of its education was hardly unfounded. From the time of colonization, Iowa had been an American education leader. In 1897 Iowa had the highest literacy rate in the nation, hovering around 99.5 percent.[8] For nearly a century the state continued to perform among the very best in the nation on a variety of education metrics. As recently as 1992, Iowa was first in the nation in both mathematics and reading in the K-12 population.[9] The state's heritage of educational excellence is reflected in the fact that Iowa continues to largely control the standardized testing market. ACT, which, among other assessments, produces the famous national standardized test for college-bound seniors, is headquartered in Iowa. Pearson, another major national testing company, is also based in the state.

However, since 1992 Iowa has dropped precipitously in national education rankings. The state retains overall education rankings in the top quintile according to most sources. However, this is likely to fall in the near future due to multiple factors. Some ranking systems already note

Iowa as performing well below the top quintile. For example, 2016 ALEC ratings place the state thirty-first of fifty in K-12 education.[10] In science education specifically, Iowa's current ranking is thirty-eighth of fifty. Most disturbingly, a 2012 Harvard study ranked Iowa last in the nation in terms of education growth.[11]

When I would talk with people in Iowa about these numbers, I got big reactions. People who didn't know about them didn't want to believe them, and people who did know about the problem expressed profound anger, frustration, and sadness. Iowa's educational decline is a source of great consternation for the state as a whole. Over the last twenty years, many voices in the state have blamed their educational declines on the changing student population. Iowa's population has become less homogenously white, and now includes many students who are not native English speakers. However, it has become unavoidable to note that many other states have similarly changing populations, and have been better able to respond to their students' educational needs. Iowa's student population remains approximately eighty percent white.[12] When analyzed as a separate demographic, Iowa's white students no longer rank among the top in the nation. While some people are still inclined to blame "the other" for Iowa's education problems, and I frequently encountered complaints about changing demographics on the ground across the state, this excuse has worn thin for the vast majority of the state's stakeholders. Local and regional education conferences in 2014–2015 often addressed these racist misconceptions directly and openly in an attempt to dispel them. I heard considerable audience resistance to this message.

Blaming outsiders is always a popular tactic for dealing with social problems but this strategy rarely if ever provides useful solutions. When we look at the facts, changing demographics aren't the big problem behind Iowa's educational declines. Continual budget cuts to education are probably a more substantial contributor. As of 2017, these cuts continue to occur. In the common best-case scenario, when cuts are not made, funding increases remain below the rate of inflation.

The state is willing to make an appearance of investment in education. For example, Iowa adopted the NGSS in 2008,[13] which mandate the teaching of evolution and climate change in K-12 science classrooms. I have spoken to many teachers in many school districts about the adoption of these standards. The majority of teachers had positive feelings about the general content and approach described in the standards but had little hope regarding their successful implementation.

The NGSS requires that science be taught as science is practiced: that students engage in the scientific method with plenty of hands-on work and opportunities to question, as is described in Chap. 2. Engaging in science as science is practiced is an excellent approach to science education, but not one that is familiar to all educators from their own personal experiences or their training. It is also an approach that is much more expensive than lecture-style lessons. Teaching science as science is practiced requires both more consumable materials and more scientific equipment in the classroom.

When Iowa adopted the NGSS, no additional funds were allocated for professional development for science teachers to help them learn how to implement the standards. No additional funds were allocated for teachers to purchase supplies for their students as their students developed questions and research interests. Virtually every teacher I have spoken with across the state of Iowa bought hundreds of dollars of supplies for their classroom out of pocket every year before the NGSS were adopted. It appears that the current funding model will ask them to give more, and then blame these same teachers for failing to meet the standards.

From the numbers in the national rankings, and from talking with teachers about the state of affairs on the ground, it has to be said that things did not look great for science education in Iowa when I began my project. Perhaps I should have been more concerned. It is entirely possible that a reasonable person would have been more concerned. But coming from the background I did, of working with Iowa teachers and Iowa students, I found it impossible to feel anything approaching despair. The teachers I had worked with all across the state were remarkably dedicated individuals who cared deeply about their students' learning. They were meeting incredible challenges in regard to funding, and in many cases also in regard to administrative support. Almost without exception, they had not given up. The students, also, I thought were a strong group. From my experiences with tens of thousands of Iowans I would describe the people of Iowa as bright, hardworking, and inclined to collaboration.

My experiences with the teachers and in the schools of Iowa, despite the challenges they faced, had not diminished the happiness I felt at the thought of my children being educated in the state. This state had historical educational superiority and it did not seem impossible, considering the quality of its people, that it should become an educational leader again. I would need to work to more fully understand what had brought the state down in the rankings and what could be done to bring it back up.

And I had the feeling I had reached the limit as to what answers I might find on the internet or in books, and that I was very fortunate to have a strong background of knowledge from hearing the stories of teachers on the ground for the past several years. It was time for a whole new kind of fieldwork. I had some ideas of what problems teachers faced. Now I would try to do something to help them. What was keeping them from teaching evolution? What was keeping them from teaching climate change? It was my job to find out and to try to be part of the solution.

MAY–JULY 2015: NETWORK ANALYSIS

In mid-April it was time for me to officially begin work on the Science Booster Club (SBC) Project. My contract stated that I had to relocate to California in July. I had agreed to this for two reasons. The first being that I desperately needed the job and would have been willing to remove several of my less necessary internal organs were that a condition of the contract. The second being that I hoped my legal troubles would be resolved by that point and I would be able to go. But, as my children's father had told me he would make it as difficult and painful as possible for me to divorce him, I did not think it was a good idea to have any sort of confidence in the smooth resolution of the case. I had to live day to day.

What a problem I faced, though. I could not leave Johnson County, a very nice county in the state of Iowa, while the divorce was pending. I mean, I was allowed outside its borders; I wasn't tracked with an implant or anything but it was not legally possible for me to move. Employment opportunities for people with my skillset existed in Johnson County, but not many of them, and I hadn't had any success getting a local job that would pay a wage that would support my family. The frank truth of the matter was that, in the local culture, mothers of my educational level and social class often did not work, particularly not full time. Many people in my neighborhood were scandalized that I would consider putting my young children into full-time childcare. Many people who interviewed me for local jobs had quite visible reactions when they discovered I had young children. The rate of marriage dissolution in Johnson County was 1.2 per 1000 in 2014.[14] This is far below the 2014 national rate, of 6.9 per 1000.[15] The fact that mothers of young children often do not work and that it is difficult for mothers of young children to find employment that will support their families may contribute to this extremely low divorce rate.

In 2015 in the neighborhood where I lived, which was full of young families, my domestic condition was rather scandalous. Divorce was frightening and frowned upon. I was raised to think very negatively of divorce, and felt deeply ashamed of my situation. Many people asked me why I could not work things out. I had several women in my extended social circle tell me that their husbands abused them in various ways, smacked them around regularly, and so on, but that they would never get a divorce. They would never be a home wrecker. I was asked how I could care so little about my children. My behavior was clearly considered quite shocking. It was a clear social expectation that I should subsume myself, consume myself, for the sake of the family. The prevailing view was that it was better for my children to live in a home where abuse regularly occurred than a home without a father. That a father had a right to treat his family as he liked. That if I could find a way to behave more submissively, if I could be in better accord with my husband, then his behavior would improve, and that, ultimately, the crimes in my home were my fault.

My desire to enact this strategy, to perhaps become a better wife and thus preserve my children and myself from increasingly cruel and bizarre treatment, had already driven me into the arms of religion. On this topic I had some academic knowledge and great personal ignorance. I had hoped to find the resources necessary to become more compliant, more tolerant of my situation. But that was not what happened.

Although I had hoped to learn how to lie down, I found a community that lifted me up. When I told the rabbi what was happening to me, he told me to get a divorce. The ways I had been mistreated had been recognized as grounds for divorce in mainstream rabbinic Judaism for about a thousand years. I was told that the primary value of a home was not that it be intact but that it have peace. I was told that I was a person who mattered. It was the relentless support of my friends in the Jewish community that enabled me to reach out to domestic violence resources, to learn how to document what was happening to myself and to my children, to call the police, and finally, to go before the judge to seek an order of protection.

During the divorce, the only place where I felt I was not expected to be ashamed of myself, where I felt like a person, was with my religious community. Nobody in my religious community required that I present my grounds to avoid censure. I faced no public trial there. At my synagogue I was just a person, rather than a failed wife. It helped me a great deal to be able to see myself in that way. Without the ability to retreat every Shabbat to a place where I was a person, where I was seen and treated as a person and not a vessel, I do not think I would have made it.

And I had to make it. I had to stay at least mostly sane. Now that I had a job that could support my family, the important thing was keeping it. The special bonus challenge: to make it clear that my work on the ground, my work here, made my presence in the region indispensable. At least for a little while. I was supposed to be the Director of Community Organizing. I had better get some organizing done. I had to pull out all the stops.

Luck (and time) was on my side. My connections from my last position were still just fresh enough to have some pull. I'd left full-time work to care for my children after my second baby was born. But people remembered me. I was able to get appointments all through the university system within the first week in the Departments of Education and Biology. I found a valuable cache of Banana Republic clothing at a consignment store, so I would not look poor enough to be suspicious.

The Department of Biology was enormously helpful in that first month. The Department of Education, very interestingly less so. When I spoke to my former colleagues in education about my project to build community support for teachers, they were enthusiastic until I mentioned evolution. And then most especially until after they saw the NCSE website. It took me a while to put two and two together. To realize that the openly creationist teacher who had won a state teaching award had been nominated by one of the same individuals who was stonewalling me. The academic censorship I had seen in the public school system had support at very high levels.

I utilized contacts through this individual to try to reach the Iowa City Community School District. Cooperation with the local school district was seen as an essential part of my program by both myself and my boss. But even as I made great strides in connecting with other parts of the community, contact with the school district continued to elude me. It took me, on average, nearly a business week to manage to get people on the phone. Then they would turn me down. Finally I succeeded in getting a meeting with their financial officer in their offices. My hope was to demonstrate to them how exquisitely nonthreatening I was, as well as to demonstrate my general benevolence and genuine 501(c)(3) status. My goals: to show them that I would in fact be able to give their teachers free money to buy science equipment and get them to describe a plan by which they might accept my free money.

I was told, in no uncertain terms, that they had no particular interest in my free money, that it was not possible for me to give teachers individual grants in their district, that I could not legally give any of their teachers anything worth more than three dollars without going through them, and that they were generally suspicious of me. I was seen as a spy, an agent of outside interests. Based in California of all places, which was definitely not a part of the Midwest. I was free to do whatever stupid thing I liked and they were free to inevitably refuse me.

This was not the reaction I had expected. I personally love free money, as I assume so should every right-thinking person. But I managed to keep my face arranged in a pleasant way and act as if this clear refusal definitely for sure kept a door open, and certainly they'd be hearing from me again! I didn't even let myself show any upset in the parking lot, in case they were watching me on the security cameras. I had been watched by cameras in my home, as part of an increasingly controlling pattern of behavior, for years. I was practiced in looking away from them and practiced enough to know when I couldn't afford to look away.

But even if I hid my face inside me, of course I was upset. Not only were my feelings hurt personally, but if I could not make these connections, if I could not firmly and very swiftly establish my organization on the ground right here, there were all sorts of magnified personal difficulties in store for me. I thought about calling my boss. I wanted help. But what was I going to say? How could I spin this significant defeat into something positive? What could I do?

Of course I felt like giving up, but then I remembered the extent to which giving up wasn't an option—what I needed to do was firmly and swiftly establish this organization on the ground right here. So I went home and I made some more appointments.

In my first three weeks I booked over fifty appointments. For these appointments I had three acceptable business outfits. Did people figure this out? Probably. I'm pretty sure some people did. Who knows? Maybe it helped. Regardless, by the end of July I had established the key nodes in my network that would allow my organization to grow and thrive. And part of what helped me make my decisions as to what appointments I would book and what relationships I would encourage was the modern science of social network analysis.

I spent some time looking at network analysis papers to learn how networks form and what makes a strong network.[16] I learned how important it is for each node in the network to have multiple links, and that while a hub network (where people are all linked to an individual at the center) is not a bad thing, it's only a stage in the development of a truly powerful network.[17] Ideally, you want to grow a network where you have more of a tight ring than a hub—a small group of individuals connected to each other, to whom many other people are tied. And then, as the network continues to grow, after a certain point you don't focus so much on adding more individuals to that ring, on either the inner or outer edge. You focus on adding new rings.

These new rings, ideally, are connected through multiple points at the inner members. You can fit a lot of rings together closely this way if you think about it. Structurally, you can form a lot of different many-sided shapes. And there is no one person at the center of that shape. The shape is held together by many strong connections between the rings.

What I needed to build was not a hub network, a simple but strong network where I was at the center. That kind of network had an ultimately limited capacity for growth, because just one person can't do everything. Instead, I needed to build a ring network.

To do that, I needed to think about more than just my relationships with people. I needed to think about their relationships with each other. I wanted to build a strong ring network and get things done fast. If I could find people with existing relationships and connect them in interesting ways, I might be able to get a strong ring going pretty quickly.

A single ring network doesn't need that many people—maybe four or five. I was just one of those central characters. It would be good if the other people in the ring had different types of social networks to draw on, and if at least some of them had existing relationships with each other. I thought about what I needed in my network: access to resources, like money, equipment, and space. Also, access to labor, both skilled and unskilled, and, perhaps most importantly, access to knowledge—to people who could help unlock more connections.

When I took all those early meetings I worked to learn things on multiple levels. The first thing I tried to learn was what I could do for the person across from me. The second was what they could do for me directly. The third involved some intuition. Where would they fit in a ring? Did they have the potential for structural centrality?

A structurally central person is not necessarily the most powerful person. A really good secretary, for example, can be more structurally central to an organization than a CEO. Building on that example, think of what kind of skills a really good secretary has: an ability to anticipate other people's needs, for one; really strong organizational skills, for another; and, at least with people they consider insiders, a good attitude. Those three attributes, more than wealth, power, or local popularity, were what I was looking for as I sought out people to place in the inner ring.

I had uses for all sorts of people, though! And I was fortunate to make connections with many people who were wealthy or powerful or socially influential, and all of these people have helped me in all sorts of different and important ways. It was unusual that I would book a meeting with a person and come away without a new, mutually beneficial connection. That is because I thought about more than what I could learn about these other people. I thought about what they might want to learn about me. I worked to anticipate their needs.

When engaged in this enterprise of anticipation, I assumed that people were working to learn these same types of things about me that I wanted to learn about them—what they could do for me, what I could do for them, and if they could make some sort of serious use of me, if we might develop a structurally significant relationship. Accordingly, I spent a certain amount of time on presentation before these meetings. I wanted to be able to do everything I could to make that information really clear to the people with whom I was meeting, because I knew we had only so much time, and that the impressions I would make would be important.

To be honest, in those first few months, where everything was so crucial, I put a tremendous amount of effort into presentation. It was important to consider what both my words and my appearance would convey, and to tailor these factors to the needs of the individuals with whom I was meeting. Appearance, additionally, was not just a matter of dress. It was a matter of composure, of controlling the degree of friendliness and formality I expressed.

A good deal of my postdoctoral research had focused on how people express formality.[18,19] I discovered that in teaching and learning relationships, where we culturally expect formality from teachers, informality often increases student learning. Teachers who used a lot of informal language characteristics and body markers tended to have students who were more successful in the science classroom. As teacher formality decreased as a result of professional development, student performance increased.[20]

This was a great opportunity for me to use my research. Culturally, I think most of us have a high expectation of formality in business relationships, certainly those of us who feel like outsiders. I certainly had that cultural expectation: a deep-seated belief that the behavior I should engage in for successful business relationships should be formal to convey power and success. But what kind of a scientist would I be if I ignored my research? There was only one conclusion I could draw. What did it matter how sad I was or how terrified and exhausted? It was time to be fun.

JULY–SEPTEMBER 2015: LET'S PLAY

In July of 2015, I was supposed to be divorced and in California, but I was neither of those things. I had managed to establish a social network around the SBC project with over a hundred members, and I had events on the calendar. The sort of events NCSE had not done before: community-based events, events in nature, and events set in public contexts, such as farmers markets. Events to provide informal science education on climate change and evolution in highly public settings, in many cases outside of science-related space. I pleaded for an extension. Perhaps I might remain until winter? I alluded, delicately, to my legal difficulties, which were surely soon to be resolved. I presented, my abject terror concealed behind carefully constructed emails, an article about our project in the local paper. A stay was granted; we could talk about it when I visited the office in September.

Clearly, I needed to work like an absolute maniac before September. It had to be perfectly clear that I was needed on the ground here in Iowa before September. I needed events booked. Lots of events, going quite far out, of increasing importance, size, and complexity, to protect myself in the event of a protracted custody battle.

However, despite all my strategizing, the first event the SBC ever did was a surprise. I had been working for months on what was supposed to be our first event: a public fair at the city farmers market, with many booths featuring hands-on activities about climate change that I had developed based on my research of the topic and with the help of graduate students at the University of Iowa, most particularly Kyle McElroy. I had been advertising this event on Facebook, along with posting highly localized science content. Through this Facebook account, I received a message from a woman who wanted some science content at her church's Bible camp.

NCSE has not historically put on a lot of Bible camp programming. In fact, the whole operation had become known as something of a cranky atheist stronghold, rather than the religiously neutral organization described in our mission. More than one person I'd talked to was under the impression that NCSE was more a collection of internet trolls than an articulate voice for science. I knew this was not true, and I knew that, as an organization, we definitely needed to work on our image. We did have a definite troll problem. When I was hired in 2015, we received increasingly frequent emails that our website was not usable in a classroom context, due to the concentration of profane and discourteous anti-religious comments at the bottom of nearly every page of our content.

At the time I thought it was amusing that our first event was at a church, but my primary reaction was excitement and gratitude at the prospect of work. In retrospect, it seems both symbolic and much more significant that we were invited, so very early on, to bring science outreach into a population in which some might think science was not wanted.

So that was the very first thing the SBC did: worked with dozens and dozens of wild children one afternoon in July of 2015. We field-tested all three of the major activities we had planned for that major public event in August. The Greenhouse Experience, where people could go inside a small greenhouse and see how oxygen and carbon dioxide levels changed in a closed environment. The Heat Island Effect, where people could use temperature guns to explore the area around them, learn how different materials absorbed heat, and then learn how these factors impacted living things. And Living Things, where people could plant seeds to take home and learn about how the different factors living things required for survival could be impacted by climate change.

NCSE had done tabling at conferences and other such events before. That was not the sort of thing I was doing. I wanted people who participated in a booster club event to have fun. Showing people a poster of dismal climate science predictions was not the level of engagement I desired. People needed things to do with their hands. People needed to do things that at least approached play. Our field test at the Bible camp assured me and my volunteers that people could learn about climate change while having fun. All the participants enjoyed themselves, at least most of the kids seemed to learn something, nobody stole our equipment, and we learned that we needed much more time than it said on the box to assemble the greenhouse.

These important lessons learned, I was free to freak out about the big upcoming public event. I had never put on a public event for hundreds of people before! The Bible camp had about a hundred children but they were a captive audience. For our climate change event, I had to draw a crowd. I needed advertisements, I needed swag, and I needed to coordinate everything related to facilities and transportation. We hadn't put together visual displays for the trial run at the Bible camp: no posters, no visual teaching tools. I needed to get my hands on all these kinds of materials, preferably at no cost. I was going all around town begging people for goods and services. I got the venue space without having to pay and I got the tables without having to pay. I begged physical posters and audiovisual equipment. I organized nearly a hundred hours of volunteer labor. I made grandiose claims so as to receive free trials of very expensive scientific equipment that I had absolutely no intention of purchasing. I coordinated with the home office to design and order swag, which meant designing a logo. This required seemingly endless revision and drama, and the many minute differences in the many versions made almost no sense to me, but were clearly very important to everyone else involved and certainly resulted in a much better finer product than I ever could have made.

I was in a state of near-constant panic. And while I was doing this I continued to take many meetings, to be calm and charming and assure my many partners that of course everything was under control and their many valuable donations would not be damaged or stolen, and that there would be an audience. Of course! There would be an audience!

Ten days before the event my children's daycare was shut down by the Department of Human Services.

I received a phone call from an unknown number early one afternoon. The social worker at the other end of the line told me I had to come and pick up my children immediately. I didn't even take the time to cancel my appointments.

I had so many projects going at this time. I had just managed to officially affiliate with the University of Iowa. Finally I could use a proper research library again. Except for that interruption, I'd had extraordinary access to world-class resources since I was fourteen years old. Being cut off from a university library had felt something like an amputation. Now my arm was complete again and I had access to an Institutional Review Board: an IRB. An IRB was necessary to review any and all research to be done involving ethically sensitive elements, such as the surveying of human subjects, if it was desired that the results should ever be published in an academic journal. The University of Iowa IRB was one of the most infamously difficult IRBs in the country to get through, known for its attention to detail and rigor, but that could be seen as a positive as well as a negative, right? The approval of the University of Iowa IRB could add legitimacy to my work if I could manage to earn a human subjects exemption.

There was that project: the gargantuan assemblage of paperwork needed to obtain academic-standard permissions in order to survey people at public events. Such a lot of paperwork to do something that it was almost inconceivable to imagine it should harm anybody! But it had to be taken seriously. Every angle had to be considered and planned for.

There was this project. I was negotiating my first book deal.

There was the August event and all the future possibilities that depended on its success.

There was the extraordinary need to maintain momentum, to grow my organization, and now I had no childcare.

The full impact of that realization took a while to sink in. My initial reaction was pure fear. What had happened? What had happened to my children?

Like many people, I could not afford to put my children in a licensed daycare facility. I had extraordinary legal bills, had only just found full-time employment, and was receiving no child support. Their father had done nothing to contribute to childcare costs or arrangements. And why should he? No court had compelled him to do so. And putting another weight around my neck might sink me, which he appeared to desire much more than his children's security or well-being.

Like many people, I did not like the place I could afford to put my children while I worked. The woman in charge, the woman whose home it was, was a nice woman. The children all seemed to like her. There were too many children. They were fed bad food and there were always televisions on in every room. When I came to pick them up I would often find my oldest, who was three at the time, hiding in whatever corner was quietest that day. How horrible I felt.

My daughter, a beautiful, sweet-tempered baby, had the same coloring as the woman in charge and had endeared herself to this woman, and in this way was kept with her and away from her assistant.

Her assistant did not seem well-liked by the children. For good reason. I learned that day that the assistant had broken a baby's arm. A baby just about my daughter's age: fourteen months to her sixteen. Spiral fracture. The kind you get from twisting.

There were two social workers in the home when I picked up my children that day, a man and a woman. The woman took down my contact information, asked if I'd ever seen any marks on my children, and said they would call me as the investigation progressed. They said they would call me when the investigation was closed and I could return the children to the daycare.

They never did call about anything. The woman who had cared for my children called and texted me multiple times, asking me when I would bring the children back, saying that things would be better. That there would be fewer children and that she had gotten rid of her assistant.

I couldn't put them back there. But what was I going to do? My friends in town helped me care for the children so I could work at least a few hours a day, take at least a few meetings, until I found another childcare solution. I was scared to take my children to another unlicensed place. Their father wasn't providing any help or any solutions but I was certain that if what had happened at their daycare got out he would use it against me in the divorce. I was fortunate that he never did ask why I moved the children to a different provider. It's not as if he ever picked them up or dropped them off.

The guilt I felt about this incident was overwhelming. Any judge, I felt, would take away my custody. I couldn't risk anything like this happening again. I was certain I would be held fully responsible. And I did feel fully responsible for anything that might happen to them. I had no family to help me with childcare and I didn't have any friends who could watch them on more than a temporary basis. There was really no option but to put them in a certified childcare center. I couldn't afford it, there was no one to help me afford it, but what was the cost compared to the thought of losing them?

After many, many calls, I finally found a center that could take the children. The cost seemed astronomical, nearly two thousand dollars a month. I asked around. No, not outrageous, was the consensus. That was what full-time care for two little children cost in this area—on an annual basis, just about the same as the median pre-tax income for a woman my age. Between daycare and the mortgage, my income would be almost completely consumed. No money would be left to pay my legal fees. We were already living on beans and eggs. But I had to do it.

The new daycare was so much nicer. I felt better working with the children there, less guilty all the time. I took on additional contract work, I sold some things, but even so I had to ask for financial assistance to pay my legal bills. One of my relatives in California helped me. I was very ashamed. I was lucky that I could get help.

The event went off without a hitch. We had about three hundred and fifty people there. People had a great time. We discovered that public literacy regarding climate change was extremely low. Perhaps a third of adults had an even extremely basic understanding of the greenhouse effect. The majority of people we engaged with at the event had never been in a greenhouse. The phrase makes almost intuitive sense to people who have visited a conservatory, or bought plants at a nursery, but not to people who have never been in such an environment.

This was an important discovery. The kinds of experiences I grew up with—the kinds of experiences many science educators assume everyone has had—were not as prevalent in the population as I had assumed. Not even at this farmers market setting, which had a primarily white, middle-class population. Not even in Iowa City, which is one of the more highly educated cities in the state of Iowa.

Our first event succeeded in many ways. Our programming was engaging, and people were learning from it. Once I got IRB approval, I knew I could demonstrate with some numbers just how much people were learning. But we did now know, before we ever did any survey work, that our programming was aimed too high. We had assumed a certain amount of common knowledge about climate change that was just not present in society. We had to be able to reach people who had no education at all about this topic, and do it in a way that was respectful and meaningful. It wasn't a matter of aiming lower; it was a matter of reaching people where they were. Now we had a better idea where to place our target.

There was another aspect I needed to fine-tune as well. I had thought this event would work as a fundraiser, that I'd probably clear over a thousand dollars through donations and selling swag. Instead, I cleared fifty-two dollars and forty-eight cents.

I'd intended to use donations to buy things for science teachers. I thought people would be eager to help science teachers, especially if they got nice t-shirts out of the deal. But it turned out this was not the case at all. People were annoyed to be asked to contribute more money to the schools. They felt they paid enough in taxes. They felt the schools wasted their money. The people were under economic pressure. They did not think things would get better.

Clearly I needed to work on my messaging. I was disappointed, but there were positive things to focus on. My first major event, and everything else had worked! Expensive donated equipment, definitely not damaged or stolen! My professional partners, seeing so many people in the crowd at a public science event, were surprised and excited. In a university town, there were many attempts at outreach. It was notoriously difficult to actually get people to come to anything, and no one wanted to alienate the science-curious by discussing controversial topics. But here people were, and at a climate change event!

That Monday I had a meeting with the outreach director at the natural history museum. She'd talked to people who had been at the event. Everyone was so pleased. Hundreds of people learning about climate change, and no complaints? Maybe this was a safe topic to talk about. Maybe my project could work, after all. She offered me the chance to contribute to the museum's annual Halloween event, when the museum opened its doors at night. They expected well over a thousand visitors.

I couldn't believe it. What an opportunity! This next event was only six weeks away, but I knew we could do it. And I was very glad to bring this news with me on my first business trip to the home offices in Oakland.

September 2015: A Strategic Retreat

How strange it was for me to be outside of my home environment. Urban density was a constant shock to my senses after five years in Iowa. I had been out to Oakland before, for the interview, but not since. And for the interview it was not as if I had time to take anything in. All my focus was on presentation. When I was being that way, being perfect, it was always in many ways as if I wasn't there.

Now I had been working with these people for four months. I still lived in constant fear of losing my job but in many ways my life was more stable. I wasn't worried about paying the mortgage, I had stable childcare and there was a certain amount of hope that I might soon be divorced. After numerous and extremely expensive negotiations a settlement had been drawn up.

I was only in town for two nights. We spent one day at the office and another at a strategic retreat at my boss Ann's home. We mostly talked about what strategies we might use to increase web traffic for our organization. My boss and I talked about plans for fall expansion for the booster clubs in Iowa, and we began to lay out a possible plan for going big in the region. If I could get more clubs off the ground, Ann felt that there was a reasonable case for keeping me in Iowa for another year and that this would be a cost-effective solution for everybody. The horrifying reality of daycare costs for two children had become apparent not just to me but also to my organization. Who would want to pay me enough to support my family in the Bay Area, where both housing and childcare costs were astronomical compared to the Midwest? It cost enough to support me already.

The importance of field-testing nationally scalable programs outside of the Bay Area had also become clear. It had become apparent that I would not have been able to do this work from a distance. A highly networked local was necessary for this type of grassroots community work. If we wanted the clubs to expand nationally, we would need to be able to find and attract local leaders. I could learn more about what qualities to look for and how to make clubs grow on the ground in Iowa than I could through a computer in Oakland. We agreed that it would be wise to keep me in Iowa for a while longer. Perhaps another year. Maybe two.

What a relief it was to have this breathing space. The chance for my program to really put down some roots was so valuable. I was very grateful that my boss and the board had decided they were willing to really give the program a chance to work in Iowa. I would be running it as a pilot study, now with IRB approval.

This important development marked a crucial stage in the formal alliance between the University of Iowa and NCSE. IRB approval of our survey work meant that we would have a chance to publish our survey results in academic journals. Right now, without university partnership, I could only ethically seek to publish our findings in the popular press to general audiences, as program improvement work. Although our organization did high-quality survey work, it was poorly placed to influence policy. We needed to reach a more specialized audience. We needed the legitimacy generated by more original work in academic journals. The seal of approval that UIowa's IRB could give us would help the SBC program to move to a different level.

In retrospect, this strategic alliance between our organization and a major research university played a larger role in the growth of the SBC program and the direction of NCSE than anything else we talked about as a larger group in that business meeting. We were not a news site, and simple attempts to increase web traffic had done little to increase our membership. Community outreach programs, both my work with the SBC and NCSE Teach, our teacher-facing program, would become a larger and growing part of the organization from that meeting onward. We were an organization founded to promote and defend science education. In some ways, we had fallen away from that core mission as the battle around science education in America changed.

At one time, legally actionable challenges around science education were common. Teachers were frequently disciplined or fired for teaching evolution, and parents sued school districts for teaching evolution to their children. NCSE was an important agent in responding to these threats. However, with the Kitzmiller trial the legal landscape changed.

NCSE was intimately involved in this trial, the outcome of which set national precedent for science education. After Kitzmiller, Intelligent

Design (ID) and related theories were officially not science. The submission of ID bills virtually ceased and such bills that were introduced rarely made progress at the state level. However, there are still legal threats to evolution, which change and evolve in response to the environment.[21]

Even as bills openly promoting creationism died in session and overt legal threats to teachers diminished or changed, covert social threats remained a serious issue. The Kitzmiller victory drove threats to science education underground. Despite the fact that evolution was included in most states' science curriculum, nearly sixty percent of science teachers avoided, minimized, or hedged their teaching of the topic in 2010.[22]

NCSE was still positioned to respond to flare-ups but we didn't have the tools to deal with these covert, smoldering threats. Why weren't teachers teaching evolution? We wanted to know, and we wanted to help them. Accordingly, when she came on as the new executive director in 2013, Ann Reid invested heavily in outreach. Her leadership developed programs that reached out to teachers and scientists. By choosing to invest in me and my work, she would also invest in programs that could impact communities.

In September of 2015, it was becoming clear that these outreach programs—NCSE Teach, the Scientist in the Classroom Program, and the SBC Project—were active and growing. While we still needed to be able to respond to flare-ups, we also needed to acknowledge that the landscape was changing. We would devote some serious resources over the next year to evaluating and developing these programs, to see if they really had the possibility to provide scalable solutions.

I should have felt great after these meetings. In a professional sense I did feel great. I knew that I had built something worth investing in, and even if some people remained skeptical, the woman who mattered was not.

But of course that wasn't the whole story. While I was in California, attempting to go to these various strategic meetings, I received dozens of urgent phone calls from my family and my lawyer. Negotiations fell apart. My ex would not sign the settlement. It was a good thing I'd been granted some extra time in Iowa. I'd have to go to court for my children in October.

For the date of the hearing my ex requested the same day that my organization was booked for our first very large event, our collaboration with the museum for Halloween. I told my lawyer I couldn't do it. We had to move the date. He filed the paperwork. We were assigned a new court date: my ex's birthday.

Was this auspicious or not? I definitely did not feel good about it. I was terrified to go to court. I didn't want to be in the same room as that man. He was banking on that. His lawyer suggested that we could continue negotiations. That I could sign a new settlement rather than go to court. An agreement that would legally restrain my movements, so that I could never leave Johnson County, Iowa, without giving up custody of my children.

But I wasn't that scared or that stupid.

I'd go to court in November. And by then, I needed to have the groundwork laid to establish three new clubs.

OCTOBER–DECEMBER 2015: EXPANSION

I was racing against the clock. The winter holidays were swiftly approaching. If I didn't get these new clubs up and running by early December, I knew I wouldn't have a shot at getting them off the ground before late February. Iowa City, with its economy so closely tied to the University of Iowa, more or less took a month off for winter break. Other communities in the region took it pretty easy, too, and of course the school districts were not interested in starting up new projects around Christmastime.

Christmas to me was not super-interesting. But Halloween (and our approaching very large event) was definitely interesting. Not only was this event important in and of itself, I envisioned it also serving as an effective advertisement for our organization. People came to the Campus Creepy Crawl from in and out of Iowa City, so this was a great chance to give people in the immediate region a good impression of the SBC program. People who attended the event weren't the only people I'd be able to reach. If I were able to show pictures and video from the event, other venues would see that we were capable of providing quality content at this scale, which would help with future bookings. And local people, people who could become part of our general audience, would be able to see this visual content and know that SBC events were a good place to go to have some fun.

My graduate student volunteers and I spent a fair amount of time thinking about what kind of content we wanted to provide. I felt it was important to present information on climate change, but the students really wanted a chance to put together something about evolution as well. After talking with our partners at the museum we were able to secure enough space to do both.

A Halloween event called for creepy content. The graduate student interns put together an interactive exhibit about parasite evolution, "Real Zombies", complete with a truly disgusting and disturbing video compilation of host–parasite interactions. And by interactions I mostly mean parasites busting out of host body cavities. It was super gross. I knew immediately that it would be very popular.

I was interested in the chance to go full-horror on climate change. So often when we teach about climate change in public we are told, as educators, to soften the message. Focus on the positive, on what positive changes we can make; try not to freak people out. But this was Halloween—freaking people out was my goal. With my interns, I developed "Eyes on the Rise", an interactive exhibit about sea level rise. I built little models of significant coastal monuments, cities, and cultural objects that people could dump water on to see how they would be affected by sea level rise. I also looped a video presentation projecting global coastline changes in a worst-case scenario.

But this preparation for the event wasn't the only work we needed to do to lay the groundwork for the expansion. We didn't want the SBCs to only provide great, possibly horrifying content. We also wanted to show that we were a force for good in the community. I emailed a bunch of teachers to solicit requests for our first grant cycle. We hadn't raised a lot of money at our first fundraiser, but I had gotten a substantial check from Rockwell Collins, a very generous STEM corporation with a branch in our area. There was enough money to buy equipment for two or three teachers. I could then highlight these gifts and their impact on students to show other communities what we could do for their students and teachers.

Getting teachers to ask me for free money was much more difficult than I had anticipated. I sent out emails to dozens of teachers. None of them wrote back! I ended up having to call teachers with whom I was vaguely acquainted and beg them to let me buy them things. After three weeks of implorations, I received a few half-hearted requests. It was the most creepy-uncle experience of my life, but if that was my new role, so be it. I had enough money to fulfill all of these teachers' barely interested desires. I put in the orders and contacted the teachers to arrange classroom visits to drop off the goods, then got back to work on the exhibition. The equipment I ordered to fulfill the grants would come in right after Halloween. Nice timing for expansion.

Putting together the planned exhibits was a lot of fun. We all put in a great deal of effort on everything from the displays to our costumes. In an effort to tap into 90s nostalgia I went as Captain Planet. My willingness to publicly humiliate myself for climate change education paid off on the big day, attracting plenty of attention while putting many of our event partici- pants in a mindset to be ready for conservation themes.

Over the course of several hours we interacted with well over a thousand people. Watching how people responded to the information on climate change was very interesting to me. The majority of our adult participants, perhaps ninety percent, had never heard about the problem of sea level rise related to climate change. Maybe seventy percent of all the participants were interested in the information and wanted to learn more. Some of the people at the exhibit who were visiting from coastal regions or had relatives near the sea stopped to talk with me about their experiences related to the phenomenon. Many people, both those with and without direct experi- ence, expressed their gratitude that someone was talking about this issue where people could hear. They were glad we were giving people opportu- nities to learn about what is happening to our world.

The other thirty percent of adults did not react to the material in a warm or engaged fashion. I had the unusual experience of seeing adults physically deny information they did not want to process. Grown people walked through the exhibit with their hands literally over their eyes or literally covering their ears. They had expressions of pain and distress on their faces. One woman came up to me and yelled at me. She said it wasn't right to talk about such upsetting material in public.

All of this feedback was very important to me. I had anticipated some negative feedback. After all, I was presenting an intentionally hor- rifying lesson about climate change. The facts were freaking scary! What I had not anticipated was the amount of positive feedback, the complete public ignorance around this serious issue, or the physicality of the denial response.

Seeing the pain this information caused some of our participants made me think deeply about the reasons behind science denial. It wasn't just ignorance, like many educators think. The majority of people didn't know about this topic, but they wanted to learn more. The desire of the public to learn more about climate change has been widely demonstrated through national research.[23] In this situation, where some people so clearly could not bear to learn more, could not even bear to hear more, denial seemed to be linked to emotional pain. For some people the difficult or impossible process of reconciling scientific knowledge with their worldview proved insurmountable, at least on that occasion. Finding ways to bridge the gap, to help fuel that process of reconciliation, would be one of my driving forces as I developed future exhibitions on controversial material.

We had more than a hundred people join the club that night. In the months to come, we would gather many more members as, armed with evidence of our progress, community contributions, and numbers, I began the process of aggressive expansion.

Over the month of November I had dozens of meetings with community leaders in West Branch, Cedar Rapids, and the numerous communities within Iowa's Amana Colonies. Generally speaking, people were very interested in what we could provide. I do not think that without the ammunition from the Creepy Crawl we would have had enough evidence of our organization's potential to make these partnerships. I arranged group meetings in each of these three new areas, where we made concrete arrangements to hold many local events in the new year. And in Iowa City we continued to develop our programming. We hosted a popular community nature hike before the leaves fell, we made plans to participate in another large science festival in January, and I was asked to give a workshop on language in science teaching at the February Iowa City Darwin Days celebration.

Our calendar was filling up and we were making many new connections. Delivering the new equipment to teachers further broadened my horizons. My first grant recipients came from a wide range of schools. I bought light power tools and wood for an incredibly passionate teacher in a rural school district who was developing her own middle school engineering curriculum. I bought a microscope for a devoted and creative teacher in a posh private school. And I bought gardening equipment for a teacher in a rural school for troubled youth, a man who can only be described as a light in a dark place.

The range of educational opportunities afforded to students at these schools was astonishing. The children at the private school reminded me of students in Margaret Atwood's science fiction novels: eight years old and asking me about heat maps, telling me their dreams of working as analytic chemists, wanting to know if I'd read about the effects of casein consumption on gut microflora.

The teacher at the public school who wanted to grow her engineering playbook was passionately focused on work: on careers for her students. She wanted the kids who went to her school to be exposed to the types of work they could do in their community, so that they could grow up and stay in their community. She pushed STEM education because she knew it meant options, not just for her students' lives but for her community's future. Agriculture and manufacturing were the industries in her area and, as the threat of automation approached, her students needed higher-level skills if they were going to work on those industries twenty years from now. She wanted her students to think of themselves as people who could build, imagine, work, and do, solve practical problems in their community, and be economically productive members of their community.

The school for troubled teenagers was literally set in the shadow of a chemical plant. In the entryway there were some of those inspirational posters, the type with an inspiring word and a picture behind it. These were a little different in tone from any posters in that theme I'd ever seen before. Right next to the door, where you'd see it every day as you left the building, was a poster showing an overweight blonde teenager, with something distinctly postpartum in both her body fat distribution and the set of her eyes. In front of her image, in bold black letters, was the word "PROVIDE".

The teacher told me, cheerfully, that the graduation rate had gone from less than fifteen to over seventy percent during his time there. He told me about how he'd managed to get many of his students interested in science by setting up intensely local projects, like growing heirloom plants to sell at the farmers market and using that money to buy testing equipment to see what the chemical plants had done to their soil. He told me about how he never gave up on his students. He told me how he drove more than an hour each way back and forth to work every day.

This encounter made me realize how important the microgrant approach could be. How else would I have heard this story? How else would I be able to tell people how bad things could be for science teachers here in America? How little support they had, the very basic things they were unable to fit into the budget. With something to give, I had a wedge that could help me get into places I would otherwise never be able to access—whole aspects of American life I had never come close to experiencing.

The three teachers I gave equipment to during that first grant cycle represented very different American experiences, differences based on class. The experience gave me the chance to see how different opportunities were for the children of highly educated professionals, the children of the working middle class, and students who were, by and large, children of the working poor. Teachers in all of these contexts were actively engaged in resource acquisition. They wanted to get more of what their students needed. They wanted to bring them more opportunities. They wanted to prepare them for the kinds of futures they might expect. I admired all three of those teachers for how hard they worked, how much they cared about their students' success, and the attention they paid to individual academic success and personal meaning. And I felt sad to live in a country where we prepare children for such different futures. The teacher at the private school would never struggle to find money for wood, so her kids could actually build the things they'd imagined. And her students would never be prepared for careers in the trades. They would never be presented, as an aspiration, that they should provide. How would they ever know what that meant? They'd have as little school-based knowledge of that concept as an eight year old in a high-poverty school might of analytic chemistry.

A few days after I took that teacher his gardening equipment, I went to a regional conference of Iowa education and policy leaders. Another jarring contrast. Many participants in attendance openly blamed our state's falling test scores on "those people's children", meaning blacks, Hispanics, and immigrants. Many people so speaking, of course, saw no division between the latter two categories. The official presenters gave information demonstrating that this was not the case, that test scores for middle-class white students were part of the problem,[24] that no student cohort in Iowa, no matter how you separated out the demographics, was performing above the national average. The presenters insisted that the failure of our educational system to adapt was systemic. But I could hear, from the conversations of others in attendance, that few minds were changed. For many people it was much easier to blame the other than to consider the possibility of change. I had never heard such openly racist discussion at a professional meeting in my life. Never in academia had I heard such talk put about so casually. I found it absolutely shocking. The bravery and relevance of the organizers in presenting on these topics and meeting the criticism and argument these topics brought out was amazing.

I was not made to feel particularly welcome at this meeting. People openly referred to me as "the spy". It was known that I worked for an organization based in California, of all places—land of fruits and nuts. It was rumored I was operating in the area to force people to teach evolution.

I guess their take on this issue was, fundamentally, correct. I mean, here I am, reporting quite broadly on their activities. And I did work awfully hard to promote the teaching of evolution everywhere my programs operated. Still, though, my feelings were hurt. And now my blood was up. Those jerks thought they'd be able to drive me out. I was going to prove them wrong. They'd see me again next year.

Hopefully, by then, I prayed I'd be divorced. I went before the judge during all this business with the expansion to defend my custody of the children. It was quite the circus and a very expensive show. I had to stand in a courtroom and hear it said by my children's father's lawyer that I should not be allowed to speak and that my statements should not be allowed to be read because reports of abuse would unfairly influence the judge. Emails where I reported my children's hunger, thirst, and self-harming behavior upon returning from their father, and the fact that I had refused his demands to fill in elaborate worksheets about how I spent my time in 15-minute increments, were presented as evidence that I was not willing to be an effective co-parent.

The judge deliberated for several days. I functioned in a state of terror. But in the end, my children were left in my care. Their father's visitation was slightly altered. He was still allowed to have them overnight. The bar for child abuse in Iowa is very high. You are allowed to smack them around considerably. But the man was ordered to pay a full year of back child support. Such a harsh ruling was nearly unheard of in my area.

So I had some security as I entered the new year. I had started 2015 in fairly horrifying conditions. I would begin 2016 in a safe place. Gainfully employed, fairly secure in the custody of my children, and ready to expand my organization's influence across eastern Iowa, a region of over 10,000 square miles. An awful lot of territory in which to be an effective spy. Or a friend.

JANUARY–MARCH 2016: GROWTH

My notes from the first quarter of 2016 are an interesting flurry. I put on an event about every ten days, sometimes more often. Aggressively booking events, making social connections, and coordinating volunteers and logistics took up most of my time. During this period I also began extensive data collection at events, which meant getting my volunteers access to human subjects training, following IRB protocols for data collection, and adding careful data management and storage to my list of regular tasks.

Although I had gotten three instruments approved through the IRB, each taking about ten minutes to complete, my volunteers and I found administering all three at events to be absolutely impossible. For one thing, they took the general public much longer than ten minutes to complete. For another, most people were completely unwilling to fill out the survey that asked for personal demographic information, even if it didn't record their name.

I made the decision to prioritize data collection on science literacy, with administration of a survey on impressions of scientific topics as a secondary survey when we had collected a sufficient number of literacy surveys at an event. As we were not tracking individuals or individual growth, it was possible to stack up a pile of surveys for each event and see how many we could get through, working with one participant and one survey at a time. We found that, with this approach, we could consistently get sufficient literacy data. The second survey, which helped us learn how people felt about climate change and other topics, came in more patchily with smaller amounts of data from fewer events.

From my notes it is clear these things were happening, that I worked with thousands of people at events, made hundreds of contacts, and made new bookings every week. However, from my notes it is also clear that I felt I was standing still, that I was continually frustrated by my lack of progress, and that, perhaps, I simply did not know how to feel safe anymore. It is clear from my records that I would work continually, with great animation, for very long hours, and that then I would be sick. I had to take a morning off to sleep nearly every week, but worked late every night after I put the children to bed. Although I had every reason to feel more secure, when I look back on the progress I made that spring, it is clear that I worked myself ragged. I acted like I was escaping something.

I don't think anyone was able to see it. I get the impression people saw me as a bright and friendly person, a definite prospect. I was invited to give talks. I was interviewed on the radio. Some statements I made about gender on my blog were widely quoted in the British media, from the *Guardian* to the *Globe*—though I was most pleased to be quoted in the *Financial Times*[25] and the *Daily Mail*.[26]

Ann put me under more pressure to raise money. My program was expensive. I hadn't the slightest idea how to raise money. I was still worried she might lay me off or, worse, force me to move to Oakland before I could legally bring my children out of Iowa. Accordingly, I decided to try and raise money through several strategies, from selling tickets to lectures with wine and cheese (of the box and brick variety, wildly marked up), to collecting donations at events, to writing grants. My intern Claire Tucci wanted very badly for us to have a summer camp that summer and she persuaded me that we should write grants to fund that camp, and so that was what we did. None of us but her were particularly optimistic about our chances but grants were things you were supposed to write and then sometimes, perhaps, someone might give you some money, and that was more or less the mindset of all of us coming out of academia in those lean years. Most of us thought of grant writing as more of a prayer-wheel scenario than anything practical. The bestowing of grants was a semi-miraculous process that seemed to happen so rarely, the grace of the funded mysterious and veiled.

Some of my student interns were inclined to spend a great deal of time polishing grants, working for hours on a single paragraph. Some of them were so intimidated by the process that they never wrote any grants at all, leaving documents blank as they gnawed their nails with anxiety. In my precarious mental and emotional state I had no concerns about cranking the prayer wheel. After all, I prayed every day, three times a day, and was not concerned if anything should come of that. Writing a few small grants, well. There was little difference in that exercise, really. More publicly acceptable if anything.

I informed my students that we would take a new approach to grant writing. We would polish nothing. We would crappily slap up a great number of honest, earnest documents. There was no point in anxiety as we would not get anything. Probably no one would even read these documents. These documents would be thrown into the void!

The approach appeared to be universally soothing. The fantasy that no one was reading our work produced an increased fluidity of expression. Our productivity increased greatly. The void received a great number of honest, earnest documents.

I was called to Oakland for the spring board meeting in early March. This exercise was expensive to the organization, emotionally difficult for my children, and less than productive professionally. I went to a dinner and a lunch with the board members where we had brief social interactions. I visited a synagogue on Shabbat morning and was stunned to find a community where many of the people in attendance were my age. That was very exciting. Then I was forced to violate Shabbat when the straps of my sandals broke, the only shoes I had brought with me, and I had to buy new shoes because it seemed impossible that I should walk around the city and go before the board in bare feet. My Shabbat violations on the trip were exceedingly numerous. I felt quite ashamed of myself and I felt that it was essential I should not say anything about my distress. One day I would have the freedom to decline this sort of invitation. That was one thing I got out of the trip. The absolute resolution that one day I would be sufficiently powerful and have sufficient economic security to appropriately and consistently keep Shabbat. Until that day, I would need to learn to make more careful and thorough preparations.

The issue of Shabbat observance was one that caused me consistent challenges. Unsurprisingly, many of the public events we put on fell on Saturdays. What could I do? The best I could. My Shabbat observance guided the construction of most of the activities we presented. They avoided technology. They avoided creation or destruction. Presenters, volunteers and myself, did not need to write and did not have to make permanent changes to structures. I attempted to minimize the heaviness of loads and the complexity of setups at the sites. We no longer attempted to collect money at typical events, limiting fund-raising to specific, advertised occasions.

In many ways, my attempts to minimize my Shabbat violations were part of what gave my exhibits power and flexibility. Technology-based exhibits often fell apart in rural or remote areas. Complex setups, heavy loads, and other difficulties discourage volunteers. Consumable supplies are expensive to scale, and they make for unpleasant cleanup work. My attempts to minimize Shabbat violations were as imperfect as my Shabbat observance but the closer my efforts came to meeting my ideal the more people seemed to enjoy the exhibits.

I did not consider it a violation to simply be out in public on Shabbat. It is no violation to teach on Shabbat. It is no violation to feel joy, to play. It was worth it to try. And the closer my exhibits came to Shabbat observance, the fewer violations I committed, the less work the exhibits caused everyone involved. By attempting to keep Shabbat, I truly feel I generated an unexpected side effect in my work. My work began to embrace joy and to embrace play.

Shabbat observance in many ways takes traditionally observant people out of the world. While I do not say it was a good thing for me to work on Shabbat, I do feel that continuing to strive to keep Shabbat, to minimize my violations of Shabbat, and to do what I could to embrace the positive commandments, the joy and beauty of Shabbat, made my work more beautiful. My striving helped me to do something good for the world, something better than I would have done if, because I couldn't keep Shabbat perfectly, I gave it up in total.

Someday my Shabbat observance may take my life in a different direction. But most of us have practical concerns that mean we must make compromises in our lives. Without the service of many observant Jews throughout American history, our country would be poorer. Of course there are times when anyone is permitted to break Shabbat, but what precisely do we mean when we say that? Where are the limits? How do we balance Shabbat observance with or incorporate Shabbat observance into our lives in the world?

These questions are too big for me. All I can conclude is that there are some ways in which people who take Sunday as the Sabbath have things easier, at least in terms of scheduling.

On the Sunday of my trip to California that September there were no work events scheduled. On Sunday I had intended to visit a variety of museums, but ended up reconnecting with an old friend. She had me take the Bay Area Rapid Transit (BART) into San Francisco. How wonderful it was in that utterly unfamiliar place to see something as comforting and Midwestern as her mom's old blue van, which had ferried me around on so many occasions over the previous twenty years of my life. We laughed and talked all afternoon. We went hiking by the sea. We ate an incredible lunch at a kosher restaurant. I bought a lot of fans for my children in Chinatown, and then in the early evening we parted.

That Monday I worked in the office after attending a morning minyan. That, another opportunity unavailable to me at home. There was little about being in the office that was different from the work I could have done at home. I sat for much of the time on my computer, typing away as always. The only difference to the organization I could see was that they were able to eat lunch with me, and the enormous expense of my flights and hotel. Then I flew home on Tuesday to my children, who had been extremely upset by my absence, and took the better part of a week to soothe.

The children did enjoy the fans. And I felt better about the probability of having to live in that part of the world. I did have friends out there. There would be more cultural opportunities for us out there, and the

Jewish community was so much more robust. My children would have different and better educational opportunities. And my job might become, if anything, less challenging, as interacting with an office of largely contented human beings ground down my paranoid, struggling nature. I would have to live in a tiny apartment with my children, but we would be able to get really good sushi. Any doctor out there would see my evident signs of post-traumatic stress disorder, and I'd be sure to be prescribed medical marijuana up to my eyeballs. I'd be able to go to a regular weekday minyan. All of these things seemed nice and not too terrifying. Once the divorce was settled, things would be different. Maybe there would be some kind of peace.

We talked about that, in the office. Where I would go. Would I go to California? Or would I go to Kansas, where the man I was seeing with increasing seriousness held a pulpit. In Kansas, I would have none of the cultural opportunities present in the Bay Area. But we did need data from deep red territory. And who knew when I would get divorced? It might drag out for another year. I made the case, tentatively, for flexibility of movement. There was some interest.

Things were allowed to drag on. I was given another reprieve. We'd talk again in the fall, or maybe the winter, and find out where I'd go.

April–June, 2016

Funding and New Connections

The pace of life did not slow when I got home, but as the spring came on there were so many signs that things were going to be okay. The incredible tension under which I'd been operating began to dissipate. Thanks to all that daycare I'd paid for, I received an enormous tax refund. It was big enough that I could fulfill my promise to my kids. Like many people born and raised in landlocked areas, both of my children were obsessed with the idea of the ocean. I'd told them and told them that when things got better we would go to Florida, we would go to the ocean and they would see the real big ocean, and that's what we did.

I was concerned that I would not be able to keep two small children from drowning in the real big ocean, so I asked Mike if he would come along, too. At the time, he was a pulpit rabbi in Wichita, Kansas. We had known each other for almost a year by that point. We had a lot of mutual associates. Sometimes it felt like it was too soon to see anyone, like I should be ashamed of myself, but I liked to talk with him a lot. It was nice to have something in my life that was nice.

We had a really good time on that trip. While we were in Florida we visited a place from my family's history: Nuthouse Beach. Or, as it is more appropriately known, Pass-a-grille Beach, home of the Don CeSar, an extremely and characteristically pink hotel. My grandfather went and stayed at that hotel when it was being run as a hospital of sorts by the government, during WW2. My poor grandpa was all crazy from being shot down in New Guinea and then lost in New Guinea. All in all, a very reasonable reason to just about lose your mind. It was really peaceful and beautiful at that beach. All white sand and very very gentle, gradual lines to the horizon. It helped me feel better. It made me feel better that I could go, that I was powerful enough to fulfill my promises.

To all appearances the whole trip was a useful show of strength. My children's father had a strange response to the whole thing, particularly the involvement of another man. I would say he seemed frightened and perhaps for the first time ashamed.

Both reactions surprised me. Rabbi Michael Gilboa was not a particularly imposing figure. He was, like many men in his profession, kind, depressed, and suffering a variety of other effects from long work hours and continual harassment. Neither the children nor I was afraid of him, and at that point we were frankly afraid of most people. He and I fought pretty frequently. I was not concerned to contradict him, would even yell at him, was demanding of him, and in his company was more or less everything that was not polished. We were both of us in our professions very controlled persons. Engaged in the kind of continual presentation that cannot be entirely an act, that in fact requires a deep competence and a genuine commitment. We became great friends within a few months of meeting each other the previous summer. By the spring, we were close enough that I tended to forget how other people saw him.

My husband unexpectedly signed the divorce papers less than a month after seeing Rabbi Gilboa. Why? I don't know. The news came out of nowhere. I was free. I was free! There were no restrictions on my movement. The settlement allowed my former husband generous visitation and minimal child support. He covered the kids' insurance through his work. This was useful as NCSE couldn't give me insurance outside of California, though they did give me money to buy some. Maybe things would be okay. Things were maybe pretty stable.

Whenever my kids were at their father's, I was working. And a lot of the rest of the time I was working too. We were booking larger and larger events, quite a few with attendance in the two- to four-thousand range over April and May, for Earth Day. We had some county fairs in the summer of similar size. My volunteers and I all got comfortable handling these large crowds. And people started to notice us. We got great feedback. I was asked to write a quarterly column in the local paper on climate change, which was great. But all these events cost money, we needed money for supplies, and fortunately for us the money started coming in.

All those grants we wrote in the cold first quarter of the year? Almost every single one of them got funded. We found ourselves with thousands and thousands of dollars. What fruit Claire's request had borne! Not only did she bring the organization the relentless optimism we needed to try new things, she brought us financial hope as well. Having some money made it easier for us to put on these large events, because we could buy supplies at scale. When we were able to purchase materials in bulk, I was able to get costs down to about ten cents a head for any of our activities. And with so much money, it became completely clear that we'd be able to put on the summer camp Claire had dreamed up.

That was a whole new level of logistics, and so was our other wonderful summer opportunity: exhibiting at the Iowa State Fair. Months of networking, botherating, begging, and pleading had finally gotten us a space. I was so excited. This would be far and away the largest and most important venue where we'd ever had the chance to exhibit. And funny, too, how well it lined up with the anniversary of our first event, which had seemed so overwhelming at the time. In one year we'd gone from events where we worked with a few hundred people to events where we worked with a few tens of thousands. It was really something amazing.

I coordinated a bunch of logistics and I started writing some more grants, because it seemed like I ought to if the getting was good. I was in a pretty good mood, things being so stable. I wanted to get everything squared away nicely before my kids and I went away for two weeks in early July. I was going to take the kids down to Wichita for the first time and then we'd all come and stay at my place for the week, while Mike and I took up all the gross wall-to-wall shag carpeting in my house upon which my children had often vomited as infants. And then there'd be that camp, and then the state fair right after that. I made sure to line up all my summer events very carefully with the visitation schedule, so I wouldn't ever have to be gone while I had my kids.

The day before we were scheduled to leave for vacation I came home from the auto shop where I'd been held up unexpectedly long for an unexpected car repair, cursing the unexpected expense, to find all the kids' things from their father's place on my front porch. Their father left a note. He was going away; it was all my fault, he'd maybe call the kids sometime.

After all that he abandoned them. Twenty thousand dollars' worth of fighting, that divorce. Telling everyone he was doing it for the sake of the children. Just wanted to hurt me, that's what I think. Gone without a forwarding address, about five weeks after he signed the papers.

Things weren't stable anymore. Not in the way it'd looked like it might be. No more time when I didn't have the kids, where I could pile on the work. No more child support and I'd have to find a way to pay for their insurance. Definitely more financial challenges ahead.

Now things would have to get stable in a new way. Things were going to be hard but I was glad.

No one was going to be hurting the kids anymore. He'd been hurting them and hurting them and the state didn't care because there weren't enough marks, and I'd had to accept it. I had to keep sending them back there. I had to stop feeling it. Now it was over.

Now we were all safe.

We went on vacation.

July–August, 2016

Big Events

Our vacation was pleasant. We enjoyed our time together, the four of us. On the fourth of July, Mike asked me to marry him. I had not really wanted to get engaged so soon after the divorce being finalized. I was bowled over, though. It was, really, a perfect day. I said yes.

He did want to marry me so badly. I still had a great deal of difficulty imagining the future. I did not know what it could be like. I did not know when I would ever feel safe. I did not know what it would be like for us to all really live together.

It was a test I suppose. An act of faith. I did love him, I knew. I knew that. The children and he and I, we all loved each other like a family. Like what a family could be like. It all seemed very frightening.

We agreed that it would be a long engagement. We would not be married until June 25, 2017. Next summer.

So I wouldn't think about that for a while. I would think about work. The summer and early fall of 2016 really tested our capacities as a fledgling organization. We ran a week-long summer day camp for rural kids about evolution. The volunteer labor that required was simply astonishing. Everybody put in some long days on that one. It required making full use of basically every professional contact we'd developed so far. And it let us learn some very interesting things.

The kids we were working with were mostly not from town. The camp was held in a fairly rural school district and we advertised the camp in every rural school district within two hours of Iowa City. We had kids coming to the camp whose parents drove them an hour and a half each way. Not your typical science enrichment audience and definitely not your typical audience for a camp focused on evolution. From the very beginning we had all sorts of interesting times.

I learned a lot of information that would be useful for other people foolish enough to run a free summer camp. My first piece of advice would be not to be like me. For someone who is pretty smart about some things, I am really dumb about other people's kids. Going into this camp, I thought that fifth- and sixth-grade students would be totally mature and civilized. Definitely able to take care of themselves, and not at all likely to draw on their faces with permanent markers. Sadly, I was wrong on all of these points. Spending a week with dozens of nine- to eleven-year-old children was, to put it mildly, an extremely loud experience.

I realized the extent of my misconceptions on the first day, when I faced our initial group of forty students in the cafeteria of our partner school. Shockingly, it turned out that they were kids. I thought they would be interested in quiet journaling, but they were more interested in loudly being ninjas. More on them in a minute. First, you need a little background on how we got all those kids there.

When we offered this free camp we wanted to make sure it reached populations that would really benefit from this opportunity. In particular, we wanted to get rural kids who would be less likely to have access to similar enrichment activities. To reduce financial barriers we not only offered the camp at no cost, we also provided all the kids with lunch, snacks, and supplies. This absolutely would not have been possible without our generous donors, including the ACT Corporation, the European Society for Evolutionary Biology, and Integrated DNA Technologies (IDT). On top of significant grants from those organizations, we received incredible in-kind donations from Scheels Sporting Goods, Subway, and most particularly Blick Art Supplies, who gave us nearly a thousand dollars' worth of free art materials.

While I've previously discussed how we got the grants, it's worth covering how we received so many in-kind donations. We asked for them in a brazen manner. I and other Booster Club members wrote and called major businesses in the area and asked colleagues to keep us informed about opportunities. Once we had funding and knew the camp was really going to happen, we called local businesses that had things we wanted and asked them if they would give us anything. Some club members were initially very nervous about doing this but everyone discovered it was not very hard or scary. The worst thing that happened was people didn't give us free stuff. We learned how to deal with that pain.

Getting the money and the goods necessary for the camp was one challenge. Another was getting the kids. To reach rural populations we asked for help from rural teachers. We made a flyer about the camp and distributed it to every fourth- and fifth-grade teacher in the rural school districts outside Iowa City so that they could inform the parents. We asked the teachers to help us find the kids who could benefit from this camp the most: rural kids who loved science and who had limited access to other enrichment activities. The teachers really came through, using their specialized knowledge of their communities to help us find lots of kids we would not have otherwise been able to reach. Some of the rural school districts also connected us with home-schooled children who were interested in science and would benefit from both the educational and the social aspects of the camp.

On the first day of the camp my volunteers and I had the fruits of our labor before us—a group of primarily rural students who were socioeconomically, linguistically, and ethnically diverse. We had good, healthy food to feed them and lots of cool stuff to do. I led the kids through a discussion of the scientific method, using their journals to do individual and small group brainstorming before sharing with the whole camp. Then I asked them to move on to the topic of evolution. What words came up when they first thought of evolution?

"Lies!" shrieked one little camper. "It's lies!"

I nodded my head and wrote the suggestion on the board. We were ready to roll.

What did I do when kids offered negative, creationist reactions to evolution? I wrote them down on the board with the other kids' comments. I didn't react negatively to the kids. I didn't argue with the children or tell them they were wrong. This was an important element to setting up the desired culture in the camp.

The kids offering anti-evolution statements got to see what other children in the camp thought without being marginalized or excluded. They got to see that when most other children thought of the word "evolution", the first words that came to mind were words like "change", "dinosaurs", and "DNA".

I and other volunteers also spent time talking with kids in small groups about evolution. We were very impressed by the depth of content knowledge many of the children possessed, including that significant minority of creationist students in the camp. We had more than one kid in the camp who associated "lies" with evolution. We also had kids who primarily associated words like "Armageddon" with evolution.

"I think, the end of the world", one boy whispered to me, when discussing his concept map. "Because if things change, they only get worse."

My first reaction to such a statement was to be shocked at how dark it was. Clearly, this kid had the potential to grow up to be a really high-quality goth. But right on the heels of that, I couldn't help but think of what an interesting window it was into a worldview I would have been otherwise unable to access. By refusing to act negatively or with hostility to our creationist students, by treating them the same as kids who were accepting of evolution, we got to hear more about their emotional reactions to evolution.

For most of these students, their education about evolution had gotten very tied up with fear. They had, in most cases, deep-seated ideas that people who believed in evolution were mean or bad, that such people wanted to lie to them, and had, understandably, a child's view of lies. They didn't conceive of the creationist/evolution debate as a conflict of belief versus evidence, a debate over valid sources of knowledge, or a complex sociological issue. They saw it as truth and lies, and, as we all know, only mean people tell lies. In their minds, it seemed clear that "belief in evolution" was a cultural marker associated with outsiders and that such outsiders were probably bad people who did not share any other values with insiders.

This, of course, is not true. By putting some humanity into these children's experience of evolution education, I think we were able to start them thinking about evolution a little differently. By refusing to fight about evolution, but instead calmly informing the students what scientists and experts thought, we failed to fulfill their expectations of outsider behavior. From the perspective of the kids the volunteer camp leaders were definitely talking about evolution but for some reason they were not mean or argumentative. Instead they were nice and giving out sandwiches. Very confusing.

When we first talked about evolution with the campers I did not focus on the creationist associations. I just wrote them down with the other suggestions and then focused on what I thought would most productively drive discussion. In this case I wanted to talk about DNA.

After lunch we were going to visit one of the world's premiere producers of DNA for laboratory environments, IDT. We spent some time talking about DNA as the vehicle for evolutionary change with the children. This was a valuable strategy when working with creationist students. Most of these students had been taught arguments for refuting evolution as it related to organisms but few had been taught arguments related to genetics. Without canned responses to provide, the kids seemed interested to listen and learn.

All the children were very excited to visit the labs. Our kids went on a tour of the facility where they saw awesome giant machines, learned about various careers, and met many different types of engineers and scientists. They also got to "do real science!", a phrase many of them used gleefully. After a brief lecture where more nice, calm, professional adults talked about what DNA was and how it related to evolution, professionals at IDT led the students in extracting DNA from strawberries using basic equipment and reagents.

The kids were so excited by this exercise. They felt very important and professional to get to wear gloves and handle sterile tubes. They were allowed to bring home vials of the DNA they extracted. Some of the kids carried these vials around all week as proof that they were now "real scientists". Not only was this super-adorable, it was also very nice from an educational perspective. The kids had a discussion about DNA's crucial role in evolution and then they got to physically hold DNA. DNA wasn't a lie or an idea. DNA was a thing they could see. DNA was a thing people could make and sell. DNA was the thing behind all the jobs they saw all the nice people doing. Good jobs that paid for the good cars in the parking lot and the good clothes the people were wearing.

Kids notice that kind of thing even if we want to think they don't. One of the factors that convinced me to focus on my education as a kid was seeing the kind of lives led by people who were highly educated. I wanted that kind of a life, where I could wear interesting jewelry and have lots of sweaters and complain about only going on vacation once a year. Seeing that kind of a life, and seeing people do jobs that led to that kind of a life, helped me believe that such a life was something I could have.

The kind of career modeling we were able to do in our camp, combined with an open acceptance of evolution, is not what our pro-science community generally pictures when we think about how to deal with conflict with creationist students. However, based on my sociocultural view of science and science education, it makes perfect strategic sense. This camp gave me a great opportunity to test another major strategy in my no-fighting approach to evolution education. I wanted to show the kids what accepting science could give them.

For our second day at camp we wanted to spend more time on evolution in organisms. Our theme for the day was adaptations. We had the kids look for adaptations in extant and extinct organisms. In the morning we took them to visit the Iowa Raptor Project; a facility that rehabilitates injured predatory birds. Most of the kids knew the word "raptor" in a *Jurassic Park* context. While some were initially disappointed we weren't going to see that kind of raptor, I reminded them that the movie ended pretty poorly for most of the humans. We talked about how birds and dinosaurs are related, and how the modern animals we call raptors have some adaptations that are like the raptors from *Jurassic Park*.

The kids were divided into small groups and set out to explore the facility. They saw many types of live predatory birds: specimens who could no longer survive in the wild and who now have permanent homes at the center. The kids noticed how different groups of birds had different sizes and shapes of eyes, feathers, and claws. They talked about how these adaptations were related to the types of animals and fish they hunted, where they made their homes, and their hunting strategies.

After lunch, we applied these conversations about adaptations to extinct organisms by taking the students to the Coralville Devonian Fossil Gorge. This site has an incredible number and quality of exposed Devonian fossils. Our friendly local paleontologist, Tiffany Adrain, volunteered to give the kids a guided tour of the gorge. Iowa City is particularly fortunate to have a very warm, friendly fossil expert who has a child of her own and is very comfortable working with children.

The plan was that Tiffany would show them different fossils and have the kids talk about what sort of adaptations they might display. As luck would have it, the weather did not cooperate. It started to rain, and suddenly we were dealing with dozens of ten-year-old maniacs running around on wet razor-sharp rocks. Dreading the prospect of explaining so many potential injuries to so many potentially angry parents, we pulled the kids into the visitor center and ran a modified indoor activity.

Again, we had creationist students object to the discussion, specifically to the age of the fossils and the Earth. This time, other students responded and said that they wanted to hear from the expert.

We didn't discourage the creationist campers from speaking up, and we answered their questions appropriately and respectfully. We encouraged everybody to have good manners and we discouraged anyone from disrupting or shouting down anybody else. But the non-creationist students' appeal to authority did have an effect on all of the students. It quieted things down better than I imagine we adults could have.

Seeing the children interact with each other on the topic of evolution versus creationism was very interesting. Having adults model positive behavior was clearly important to them. The kids knew it wasn't acceptable for the situation to devolve into a shouting match and that we wouldn't tolerate anyone being put down. The message was clear: we were there to learn. All of the adult volunteers acknowledged the paleontologist's authority on this topic. When the kids asked us questions about the fossils that we couldn't answer we also asked the expert.

Demonstrating a respect for expert knowledge is an important part of evolution education, as well as admitting when one is not an expert. Some of the creationist students had genuine questions for our expert once they realized this wasn't going to be a debate about the age of the Earth. These students wanted to know how scientists knew how old the fossils really were. The students were genuinely curious and they received clear and detailed answers from an expert. Their questions were respectful. The kids seemed thoughtful.

Some of my volunteers were disheartened that we were still hearing creationist objections on the second day of camp. I thought it was a good sign that people were still talking. Many educators have had students sit silently through a lesson on evolution, then unexpectedly produce very negative course evaluations. If people are communicating and asking questions, we should always take that as evidence that we have done something right in structuring a learning experience.

We pushed the kids pretty hard on evolution for the first two days of camp. On the third day I felt it was time to pull back a little. In the morning we were going to take the kids on a nature hike to a fairly fossil-rich area where we hoped to enforce yesterday's lessons on adaptations in extinct and extant organisms. In the afternoon we would tour Iowa's State Hygienic Lab, a facility that tracks all sorts of critical public health measures for the state, where we would conduct lab experiments and meet lots of scientists. There wouldn't be any explicit focus on evolution during this time but we would be in another environment where acceptance of evolution was normal, where the kids would get to see and learn from adults with comfortable careers in the sciences, and where they could see the diversity of scientific careers in the public as well as private sector.

The weather was beautiful for hiking. We all enjoyed being outside and there were many interesting plants and animals to show the kids on the trail. We taught the kids about medicinal plants, we got to see lots of giant water snakes (always exciting), and we even sighted my personal favorite bird, the little green heron. And then, on the rock dam where the kids could see so many fossils, there was a fight! Some of the kids started arguing about the age of the fossils and the age of the Earth. One of the creationist campers became very upset and was yelling at the rest of the kids that they were wrong because the Earth was four thousand years old. Some of the other campers began yelling at this kid and calling him names.

One of our volunteer group leaders stepped in, a University of Iowa undergraduate, Jorge Moreno. He asked the kids to calm down and stop yelling. He said that everyone was allowed to think what they wanted to think, but that they should also be open to listening to people who specialize in this field and ask them questions before jumping to conclusions.

This approach worked out very well. The kids stopped yelling at each other. By taking the social pressure off the creationist kids, we reduced the insider/outsider problem that could have kept them from learning about evolution. Most people tend to double down on their positions when they're backed into a corner. Feeling trapped and under attack does not put people in a frame of mind to think or grow. And that is definitely not how we wanted any of the children participating in this camp to feel.

We finished the hike without further conflict, except for the outraged camper responses to cruel prohibitions against swimming, catching snakes, or whacking each other with sticks.

The next two days had a very explicit focus on evolution. On Thursday we went to the National Mississippi River Museum, which had a new exhibit on dinosaur evolution. On Friday we toured evolutionary biology labs at the University of Iowa and then learned about evolution in the state of Iowa at the Museum of Natural History.

When we divided into small groups on Thursday, I had the care of one of our more outspoken young creationists. I was happy about this because he was an incredibly smart boy with a sweet personality and I found him pleasant company. I brought my four-year-old son to the museum that day, and watching this boy and my son interact was really enjoyable. When my four year old had trouble keeping up with the bigger kids, this boy slowed down and held his hand, saying, "It's okay. When I was your age I thought I would never get bigger, but I did and you will too!" This is the same kid who was the first one to offer to share his water bottle with a thirsty teammate, with the explanation that they were "both sons of Adam, so it's like we're brothers anyway!"

When this boy was standing next to me at an exhibit of fossils, away from the other boys, he pointed at one.

"That's 175 million years old", he said, quietly.

I agreed with him. And we didn't say anything more.

It was very interesting how after an open conflict, which group leaders responded to calmly and by de-escalating the situation, the strong majority of our creationist campers began to show signs of accepting evolution. By working to reduce ingroup/outgroup stigma, it appeared we had made more progress on this topic than is generally seen in purely fact-based approaches.

What was particularly interesting to me was how quickly and completely the campers changed their position on evolution. There was no transition period where I heard creationist campers go back and forth between positions. We went from campers openly fighting about the age of the Earth on Wednesday to every single kid in the camp showing every sign of accepting scientific fossil dates on Thursday. Kids who had displayed a lot of fear tied up with the concept of evolution seemed less nervous. We didn't have any more fights. Kids who came into the camp expressing creationist views, on average, asked more questions at the museums on Thursday and Friday than kids who came into the camp accepting evolution.

The children all seemed engaged by the tours and exhibits at the museums. When they had the chance to tour evolutionary biology labs at the University of Iowa, they were polite to the scientists and enjoyed learning about the model organisms. We visited a lab that works on *C. elegans*, a common model organism that is often hermaphroditic and self-fertilizing.

This was explained using that vocabulary by one of the scientists giving the tour. I asked the children if they knew what "hermaphroditic and self-fertilizing" meant. Unsurprisingly, they did not. The scientist explained that the worms had male and female parts and could reproduce by themselves. An astonished camper informed me afterward that she'd thought all animals went "two by two!", a clear reference to the story of Noah's Ark.

There is no question that the camp exposed these kids to information that shook up their worldviews. This is exactly the kind of situation where one would expect a lot of conflict and negativity. Certainly a lot of parental complaints. In a blatant attempt to provoke possible parental complaints, we even gave all the campers free *Darwin Days* t-shirts, which advertised *Darwin Days*, an annual local celebration of evolution, complete with a phylogenetic tree.

The provocation did not work. The parents who were sending their kids to my free camp also appeared to want my free shirts. They were pretty nice shirts.

In our final journaling exercises I gave the kids chances to complain. I asked them to write what we should change next year, or if there was anything they didn't like this year. Almost all of the campers wanted desperately to come back next year. A smaller group demanded tacos for lunch in the future. The campers had different favorite activities but no one complained about evolution or creationism. They enjoyed the camp.

At the beginning of the camp around ten percent of children expressed creationist views. We saw no evidence of creationist views in the children's final exercises. This is frankly astonishing—and it should be to anyone who works in science education. All the literature tells terrible tales of how it is incredibly difficult to get any sort of shift in entrenched creationist populations. But here we have a sign of hope!

The culture of open, calm, accepting communication we fostered in our camp was very important. The fact that creationist campers weren't made to feel like outsiders helped them feel safe to ask questions. The fact that we refused to stigmatize any camper, and that we actively worked to include all campers, helped all our students feel like insiders.

The week was such a learning experience for everybody, not just the SBC crew and the kids but also the evolution education community. Who would have thought we would be able to make such educational gains in a creationist population? It still surprises me. Maybe it was because I liked them. That might have been part of it. Those kids were nice kids. Good kids.

And then, there was the Iowa State Fair.

The fair. I'd been dreaming about going to the fair since the option first presented itself in the early spring. What a lot of work it took to get to the fair! Here, I didn't need to use my social organizing skills. I needed to use my bureaucracy survival skills. It turned out, unsurprisingly, that some people high up on the food chain did not really want us to talk about evolution in public. Not that much in public. Too much potential for controversy. I would not have been able to get us in at the fair if not for my friend, Tiffany Adrain, who allowed me to partner with her through the Iowa Museum of Natural History. She wrote a proposal that included some essentially true information about what we were going to present but neglected to include the "e" word. Ooops!

I thought it was so brave of her and of the rest of the museum staff. They could easily have been dinged in following years, never mind the many other possible repercussions. But they wanted to talk about evolution directly in public. The chance to do so in front of such a large and diverse audience was extremely valuable to us all.[27]

You may or may not have heard of the Iowa State Fair. The Iowa State Fair is huge. Over a million people go every year, more than a hundred thousand people a day. The spot we had secured was in the Varied Industries building, which is visited by the majority of the people who visit the fair. Why? The fair is held in August, which is extremely unpleasantly hot and humid in Iowa. The Varied Industries building is one of two air-conditioned structures on the fairgrounds. We anticipated a continual river of people, especially since we would be at the UIowa booth, where fair denizens are able to obtain free college sports calendars. In a state with no major league teams, people in Iowa go nuts for college sports. The Iowa Hawkeyes, with their extraordinary athletics, were a huge draw.

I anticipated that people who enjoyed sports and people who enjoyed air conditioning would constitute a broad cross-section of society. Perhaps even the majority! And the fair provided such an excellent society from which to sample.

The state fairgrounds are located in Des Moines, the capital city of Iowa. This city is conveniently located in the center of the state. The vast majority of Iowans are less than three hours from Des Moines. About three million people live in Iowa, and every year about a third of them go to the fair. All kinds of people go: people from every town, people of every socioeconomic class, and people with every kind of interest. It doesn't matter what you like. Something you like will be at the fair.

We got the word that summer that Tiffany's proposal had been accepted. We would be permitted to exhibit for a whole day. Not even a half day, a whole day! And, that day would be Friday. For my purposes, absolutely the best possible day.

On the Friday we were to exhibit, Kiss would be playing at the main stage. There would also be the final round of a popular rodeo event where small children attempt to ride sheep. I was certain the crowd would be enormous.

And it was. It really was! We saw tens of thousands of people. The fair admitted over one hundred thousand people on the day of our exhibition, and it was a hot day. Nearly every person at the fair must have meandered through our air-conditioned exhibit hall.

It was a fascinating experience and very intense. At times, there were many dozens of people lined up to see our exhibit. As usual, we had difficulty collecting data. I was terribly upset once the day was over to discover we'd collected only sixty surveys. But in the press of the crowd it was difficult to do much of anything, let alone get people to fill out complicated papers. People were there for the temporary tattoos and to touch rare fossils, not to help me with my research study.

We were far from alone in the exhibit hall, which was crammed with over a hundred other exhibitors. One of the other booths, we were very interested to see, was staffed by our opposition: the creationists! Their booth was well staffed and very fancy. It included such highlights as a giant mural depicting giraffes feeding with sauropods and pelicans sailing with pterosaurs. The staff handed out realistic-looking fake million dollar bills to attempt to draw people into the exhibit. I collected their literature and observed their crowd while I was on break. Although they were engaged in very active carnival-barking, their crowd was nowhere near the size of ours. Our exhibit had a considerably higher rate of organic draw, and we were a much more seat-of-the-pants operation. We weren't giving out any fake money and we didn't have any highly creative wall art. I felt heartened by the evident public attraction to our reality-based exhibit.

My students spent some time interrogating the creationist staffers. They told me that these people were disseminating entirely false, wholly unscientific information about many topics, including basic genetics. In a population that was educated about even basic genetics, the arguments being presented by these creationists would have been exposed as obvious flimflam. Based on our experience at the 2016 state fair, we would go on to develop a traveling exhibit that would counter the current claims of the creationists but without engaging them directly, let alone engaging them in debate. By spreading knowledge about genetics and evolution, we knew that we would arm the populace with the knowledge and skills they needed to make their own decisions.

What a good time we all had at the fair. For me it was so calm. The student volunteers were so active and engaged that I only had to work one shift. My colleague Minda was visiting for the fair, and she and I had many discussions in many places, such as the ski lift that you can ride all the way over the fairgrounds. It was clear to me that she had been sent out to Iowa not so much to see our exhibit as to pump me for information. What did I see happening with my project? Did it have room to grow? What should the future be like for it? What kind of direction did I see for the organization as a whole? What should my future be like?

I knew enough to say something, I suppose. It was hard for me to envision the future. I wanted to have a future. I did think my programs could grow, that they could be good and do good. I wondered why Minda was having all these conversations with me.

It turned out it was because she was leaving. She got a better-paying job at a more environmentally-oriented nonprofit, the Sierra Club. While this was clearly the right career move for her, I was sad to see her go. I had relied on her a great deal as I was learning the ropes for my job, and the organization as a whole had also relied on her a great deal. She was so competent and unafraid to do things, to try new things. She had been with the organization for three or four years when she left. There wasn't any more up for her at NCSE.

When I knew Minda was leaving, the conversations took on a different angle, because it meant there was somewhere for me to go up at NCSE.

My children's father had not returned after depositing all their possessions on my porch. We had not heard from him, we had not received any child support from him, and the children's insurance through his work had lapsed. Purchasing them new, inferior insurance through the exchange was costing me five hundred dollars a month. He had been paying me seven hundred dollars a month in child support. Considering the income depletion and this new significant bill, I was under quite serious financial strain. And, while I had hated having the children go to their father's, where they were poorly fed, left dirty, and picked up skin infections, and my son was slapped around and mocked until he cried, it was a change to have them with me all the time.

For them, the change was good. The transition was stressful, of course. They were traumatized from having their parent abandon them. But they came to be secure. To know they were secure. Their health improved dramatically. They became calmer and more settled. They slept through the nights better. Their play was more like children's play. My son stopped hording food. For them, the change was good.

For me, having no days where I could focus on work was hard, and having no mornings where I could sleep in was hard. I had to find a way to work less at night, after I put them to bed. But when else was I supposed to finish my work? When could I do my side work? I had absolutely no money to purchase supplemental childcare. I had fairly little money with which to purchase food. We were living on about a hundred and twenty dollars a week, after bills, which I knew I should be grateful for, as it was much more than many people get, but I did find it difficult. Gas and diapers and food and clothes and school supplies all had to come out of that, and I had to buy enough gas to drive just about all over the state, trying to keep up appearances well enough to make the types of professional contacts I needed. My resources were badly depleted from the divorce. I needed to make repairs to my home. One of the walls had turned green on me and I felt that issue really demanded professional attention. The plumbing had become temperamental to the point of approaching sentience.

I felt bitter that I had worked so hard and done so well and had no peace or reward for it. Although I knew that I was very fortunate to be able to keep my home and keep my children safe, and not have to be on assistance and not have to sell all my jewelry, the scratching, stretched feeling of my life was painful. All this responsibility, all of it on me. No help, no end to it. Life would be like this for years and it would never be easier.

Unless I made more money. And with Minda gone, I knew there was more money. So I became vicious and intolerable in my pursuit of more money. It wasn't right that I should do so much work and do so well and live on beans and eggs; it wasn't fair. It made me angry. I got to the point where I was willing to show that I was angry. I got to the point where I was not willing to be nice at all in these negotiations for my future.

I was given a raise that increased my income by about five hundred dollars a month, after taxes. So I was not so well off as I would have been had I been receiving child support and their insurance covered by their father, but I was not stretched so thin. I ended up with a seven hundred dollar a month drop in income instead of a twelve hundred dollar a month drop in income due to his abandonment.

And, in these negotiations, which took nearly six weeks, which completely exhausted me and made me weep most evenings, I secured a different type of future at NCSE. I was to move to California and take on a supervisory role in the summer of 2017.

Did I want to go? Not really. But I did want my life to change. We had to go somewhere. My fiancé had decided to leave his pulpit. I told him about the outcome of the negotiations. He was less than thrilled at the thought of moving to the Bay Area, one of the most unaffordable locations in the United States. But there was no work for him in Iowa City. We had to go somewhere. In California, we felt we could see something like a future.

September–November 8th, 2016

Well, That Was Pleasant

My promotion secured, let it be said, I felt like a pretty fancy lady in September of 2016. I told everybody I was moving to California and contacted a realtor to put my house on the market in the spring. She told me I needed to paint just about every inch of the place. My kids were getting a great quality of care in their new schools. I had enough money I could get my toilet fixed and buy a new pair of boots. I adored buying whatever produce I wanted. Life was good.

And the great news just kept coming. My student interns and I had our most successful grant season yet, with a cash take over 13K. Not bad for a program less than two years old. I was able to start analyzing longitudinal data, which revealed a significant increase over time in adult scientific literacy in Cedar Rapids. You can read the paper if you want! The results turned out to be really interesting.[28]

The SBCs became a civic institution, building long-term partnerships with the Iowa City Public Library and the Department of Parks and Recreation. I was nominated for Division Director of Research for the National Science Teacher Association (NSTA), a glamorous and powerful position that would necessitate my staying in fancy hotels and eating at nice restaurants. I was invited to the Howard Hughes Medical Institute to share my expertise on conflict reduction around evolution, also a seriously swanky deal. I was asked to review papers for some big-name journals, and some of the stuff I reviewed cited my work—an exciting new experience for me!

Fancy, fancy pantsy. I was feeling pretty good. I hadn't been working on my ground game very much. I made the decision after I got the promotion that I had to focus on funding and stabilizing my existing operation for after my departure, instead of increasing on-the-ground expansion. Maybe I made the wrong decision. I can't say I didn't work hard. After all, I did get my operation stood up, with the resources it needed to survive and continue to grow at a modest, steady pace after my departure. My mindset shifted as I got ready to move. I became less of an on-the-ground community organizer and more of an executive. I was coordinating volunteer operations in three states, had frequent calls with people in Tennessee and Virginia who were trying to get things off the ground to do some of the things we were doing in Iowa. In an average week, I helped coordinate operations by directly talking with twenty to thirty people involved in volunteer efforts, some of whom were responsible for more people under them who I never directly touched. I also met with or contacted at least one potential major funder a week, a person or organization who could potentially give us five-figure levels of support.

And it came out that by November, counting in-kind donations, I had definitely managed to pull in five-figure support for my operations. For 2016, my cash total was 23K, my in-kind 35K, and my known volunteer hours could be valued at 17K if evaluated at the standard nonprofit rate.

This level of community buy-in, especially so quickly, had not been expected by anyone back at the home office, except possibly Ann. When I was going through my numbers, I recalled when our accountant took me to the bank to get me signing privileges on an account in Iowa. She asked me how much money I thought I was going to be running through this account. I told her I thought I'd get over ten thousand dollars in our first year. She laughed at me! Now look where I was. There are few feelings sweeter than succeeding in the face of doubt. I'd gone from nothing to a financially healthy regional nonprofit in eighteen months. I was providing a valuable service to my community, at a budget that dramatically undercut every other outreach effort that had shown me their books. And I was ready to expand.

I was fussing at the home office, trying to get them to help me make a donate page for the Booster Clubs. I wanted to hire some staff to get the ground game rolling again. We could lose our momentum on the ground if we didn't! Everyone was helping me out. Although not everyone had thought the Booster Club Project was going to turn out to be much of a thing, it was becoming clear that this was a strong area for growth.

I wanted to get a call out to our members, of which I now had 872. I wondered how many would give the club financial support. We'd never asked them before, but I had received quite a few spontaneous donations. Some of them pretty big, like $75! That was a lot of money. That's about what I spent outfitting my boy, getting him ready to go to big kid school instead of daycare—new socks and underpants, new lunchbox, second-hand clothes. That was about the power bill, $75.

It seemed plausible that if I put out a good ask, telling people how their money would directly impact our region, and showing people that we really could do the job, I might be able to get a thousand dollars out of them. That'd be enough for us to get our new exhibit on genetics and evolution up and running, and take it all over the place. Or it'd be enough to cover the material costs at about five big events, reaching at least ten thousand people. That was about the ratio, when I looked at the numbers. Every dollar let me reach about ten people with one of our exhibits.

That's the kind of thing I was fussing about on November 8, in between feeling fancy and admiring my new boots when I walked past plate glass windows. They really were nice boots: real leather, German-made, and calf-high. Broken in on the first day, perfectly comfortable to wear. Getting ready to hire more staff, advocating for Claire's promotion in the organization, as there was no way I could manage a substantial expansion without her help, getting cash into the hands of my student interns and support set up for them for spring. Feeling proud, getting high-quality new exhibits out there, reaching at least two thousand people a month. And that was in a slow month. Just completed a microgrant cycle, where we put durable goods into the hands of about eight hundred more students. Doing good. I was doing good.

I bought some strong beer to help me tolerate the election. All the polls seemed so clear. I'd been fussing and nervous about the election, I was still fussing and nervous about the election, but all the men in my life were doing that thing where they sort of soothingly patronize your concerns and it really does make you feel like everything is going to be okay. I put the kids to bed that night, I sat down in my spot on the couch, where I could see the TV, and I had my laptop and my phone on one little table right near to hand, and another little table for snacks, and it really was very cozy in the spot where I almost always spent my evenings. I put on some dumb show my friends were watching on the TV in the background, I opened the *Washington Post* on my laptop to watch the live election map, and I started chatting on my phone. Wasn't it funny? Texas was blue!

Then I saw it all turn red.

NOVEMBER 9–DECEMBER 2016

Early Response

The day after the election was a real heller. Everyone weeping, everyone wondering what the hell kind of America this was going to be. I put out a fund-raising call on my own pages, and I got NCSE to put out a fund-raising call on their own page. A response to this situation—a way for people to do something in response to this situation. Something constructive. I put out a new volunteer call on my local channels.

The response to these calls was strong. The volunteer call reached over a thousand people through social media shares, strong for a micro-network. I received hundreds of dollars from local sources. NCSE received thousands of additional donations.

My town got kind of crazy after the election.[29] There were scary anti-gay incidents. Anti-immigrant incidents, anti-Muslim incidents, anti-Semitic incidents. You name it, we had an incident. Assholes brandishing Confederate flags right across the street from my baby girl's daycare.

What were we going to do? Almost everybody I knew seemed to feel so paralyzed. What were we going to do?

I couldn't do anything but get to work. I was so upset.

I felt very fortunate that I had work I could do. I'd been proud of my programs before, I knew they could be something meaningful and important to people, but I'd never felt such dreadful seriousness. In my time with NCSE I'd become acquainted with many other organizations that did science outreach. I felt we were competitive with all of them but now, with the political map having been so dramatically redrawn, I saw so clearly what was different about what I had built. We were the only outreach program of which I was aware that was growing and thriving in the red. The new red.

As I begged desperately for money, I coupled my plea with a promise to push westward. To hit the state capital, so as to gain more political influence. And from there, to continue the push into the increasingly economically challenged western half of the state. Into the deep red territory where, so far as I knew, no one was bringing back much data on climate change or evolution. And in this changing political climate, where the Department of Energy was asked to release the names of all employees who had attended conferences on climate change, where the president-elect publicly stated that climate change was a hoax, we needed that data more than ever.

For the rest of November I put in sixty-hour weeks. Not only did the money come in, with support from dozens of individuals and several major institutions, including the University of Iowa, the European Society for Evolutionary Biology, and the National Science Foundation (NSF), but so did the people. At our monthly meeting that November, I organized more than eighty hours of volunteer labor, a new record. And volunteers began to pour in from states across the nation, wanting to start booster clubs in their own communities.

We'd begun to get volunteers since the early summer, and had people working slowly to organize themselves in Tennessee and Virginia. By the end of November, that list of states was joined by California, Colorado, Massachusetts, Maryland, Nebraska, New Jersey, New Mexico, New York, Ohio, Oregon, Pennsylvania, Rhode Island, Texas, West Virginia, and Washington. One of my interns pointed out that I really ought to begin writing things down about all of these states. When I put the list on paper, I could hardly believe what we were looking at. I had to get a map of the United States and put stars on the states where we now had a presence to really understand what had happened.

Amazing. It was amazing. If I were running a chicken franchise, I'd really be rolling in it. But, as a research scientist, contrary to some opinions, I of course was not.

I talked to one or another volunteer team almost every day. And I started digging in on my ground game, working to keep my promise to my donors. We expanded to Grinnell in early December, just like I had promised we would. And in mid-December we were able to hire Brian Pinney to work part time out of Des Moines, organizing there. I also made connections with faculty in the Department of Sustainable Agriculture up at Iowa State University (ISU), thanks to an introduction from one of my grad students to one of their grad students, who connected me to her professors after being slobbered on and harassed by my children during a weekend visit to my home.

Why did I hire Brian Pinney? He was a good pick, to be sure, with a PhD in science education, deep roots in the area, state-level education connections that rivaled or exceeded mine, and a basement that could be easily filled with tons of science equipment. But he was not my first pick. When I begged and pleaded and literally made myself sick fighting with Ann over hiring staff to help me with my expansion on the ground, I had a different man in mind—who had not only a PhD, but seventeen years of community organizing experience.

But he wouldn't take the job when I offered it to him. Through November and December, he and other influential members of my congregation warned me to be careful. They were concerned about my safety. It wasn't necessarily such a good idea for a Jew to be so public, so visible. Not when I actually went out there among the general population, in areas where anti-Semitic hate groups were known to be operating. Our rabbi received regular reports from the Southern Poverty Law Center, and had been badly frightened by the last one. We didn't need a report to be frightened. There had been a lot of incidents. Swastikas all over the place. Neo-Nazi propaganda leaflets on all the college campuses across the state. In Cedar Falls, numerous examples of public graffiti about killing the Jews, because white lives mattered.

Frankly, it was very scary. I felt almost as vulnerable as I felt called to action. I didn't stop. I wouldn't stop. But after repeated and loving warnings, I felt it would be moronic to be completely unheeding. I did have a great deal of coordinating work to do, coaching and training volunteers, and developing new partnerships. And I had a great deal of work to do pursuing funding, and a great deal of writing and exhibit development. I decided that I would spend much less time actually staffing events myself. I would go out to do quality control. I wouldn't be completely absent. But I would, as exhorted by older and wiser persons whose opinions I deeply respected, make less of an active effort to get myself killed.

I was the sole support for my children at that time. We had not heard from or received any money from their father since he abandoned them in June. At that time in their lives, and likely still now, my children were very physically beautiful. Very striking children, visibly non-white children. Not entirely typical for the area. A religious Jew, packing liberal lies to a large audience, with a visible internet presence, living alone with her mongrel children.

What an awful feeling it was. Having to take precautions. I didn't let myself feel it too much. I decided to let it be normal. Everyone in my congregation was so happy I was going to California. They were glad my plans to leave were well in place. Many people in my synagogue began making plans for just in case. Contingency plans. Should things truly turn sour.

My people. The people who had protected me and lifted me up, who had helped me to feel like a person again. Civic-minded people, significant contributors to the arts and the sciences. Servants for the public good. Charitable people, just generally very nice people. And they wrote that they should kill us all around in Cedar Falls.

It was a very sad time where sometimes I had trouble feeling anything at all. I would walk around feeling like I was wearing my body like a suit. I would perform and smile and be very appealing to people, and they would agree to almost any deal, so I knew I was performing very well. I knew the expansion was going very well. Then I would go home and take care of my children. And then sometimes in the evenings, after they had gone to bed, I would just cry and cry. Or sometimes at inappropriate times, hardly feeling anything, I would notice that I had begun to cry and cry. How awful it was. It was very awful.

By mid-December I had done everything I'd set out to do by the end of the year. I'd expanded our ground footprint in Iowa tremendously. I'd raised many thousands of dollars. I'd gotten the Genetics and Evolution exhibit off the ground and twice field-tested in small environments of twenty to forty people, so that we could be confident in executing it at a large event in January, where we anticipated serving about two thousand. And then I began to pack it in.

Everybody was getting ready for break. Getting ready for the holidays. I figured I would too. A couple of weeks with a bit less intensity. More time for writing, which I'd been neglecting. Time to shop for the children's presents. A week in Wichita doing interfaith work with my fiancé. I would buy some pretty dresses and take more baths. Maybe I would stop suddenly crying if I took a rest. I had to hit the ground running in January.

There would be many new challenges. Getting all the new sites off the ground, getting my new staff and interns trained. Finding out if I won the NSTA election. Going to Howard Hughes Medical Institute (HHMI), flying first class for the very first time. Putting out save the date cards for my wedding. It would be nice, wouldn't it? It would be very nice. I had to find some way to stop crying all the time.

JANUARY–MARCH 2017

Storm and Calm

To my surprise, I did find a way to be mostly alright, again through work. On January 24, 2017, the national launch of the SBC program was official. I sent kits to leaders in ten additional states to help them educate their communities about climate change. We hoped that half of those people would follow through and actually develop clubs. It seemed reasonable to assume a high failure rate among volunteer leaders in a beta test expansion.

Successfully recruiting leaders and launching the expansion helped me feel much better. Many of my friends and associates were depressed, angry, or dissociated in the lead-up to the inauguration and in its immediate aftermath, when we began to hear about the new cabinet picks. These included people whose stated goals included more or less destroying the organizations they had been nominated to head. On the chopping block were two topics well within NCSE's wheelhouse: public education and climate change. There were plenty of other things on there, of course— serious threats to women's rights, reproductive health care, health care in general, immigrants, both legal and illegal, police violence, freedom of the press, queer rights: kind of a you-call-it. I can absolutely understand why so many people found January of 2017 completely overwhelming, like a bad dream from which they couldn't wake. No one could take on all those issues at once. The impossibility, even, of donating to sufficient causes!

I never felt gladder to work at NCSE than during this time. While I could not do everything, I could do something to fight back for science and for public education, and that was what I was doing. The national network was small, and still immature, but it was bringing information on climate change to people in red and purple states. It was doing it in a friendly, respectful, meaningful way, a way that had been shown to work in counties all over Iowa. I knew for a fact that this national network, even if many of the leaders failed to catch, would make a difference for communities and bring in very interesting stories that I could tell on the national stage. Also, I knew that if these leaders were unable to build networks, I would still be able to find more volunteers. After we announced the national launch we had the ten volunteers we needed in a week and a half. I had to close the trial and form a waiting list.

The launch complete, my experience with first-class travel awaited. I was asked to go to HHMI to talk about community engagement with evolution education. I left Iowa on January 25, the day after the national launch. This gave me just enough time to make some really classy maps in MS Paint to help people visualize our new reach. I found an image with all fifty states and put smiley faces in states where we had volunteers and stars in states where clubs were officially running. It made for an image with content almost as impressive as its production values were poor.

I had many stimulating conversations at HHMI. The whole adventure was a very interesting experience. Of the dozens of participants, only I and one other speaker, Jamie Jenson, were not from large liberal enclaves. Interestingly, while we had never met before, we were trained in the same intellectual lineage by the same major professor: Anton Lawson, of Arizona State University. Jenson had been his last student. That is, until some years later, when I burst into Tony's office to ask him to take me on and managed to pass the numerous reasoning assessments he had developed over the course of his career. He was one of the greats.

While the other participants were highly educated experts with excellent qualifications and meaningful work in their fields, I never felt the divide in our country more clearly than I did at that HHMI workshop. It was very clear that many of the experts in education and outreach had little to no direct contact with most of their audience, that they did not understand the culture of the people they were most interested in reaching, and worse, as is entirely common with people who do not know each other, that they did not quite see them as people. I heard many negative and disparaging remarks about people of faith, people in rural communities, and, in general, people who were "other" from them—who had brought our country to this pass.

Sitting in a happy crowd of people, all well-dressed, all the ladies nicely bejeweled, the delicate but tangible feeling of contempt filled the room like perfume. The others were a problem to be solved, not people to be known. There was a serious and general conflation of ideological difference and dehumanization. Observing this contempt in a stronghold of the liberal elite, I could feel very clearly why so many people who were struggling chose to vote for Trump, even against evidence. I could see how a person's pride would cause them to vote for someone who did not radiate obvious contempt for them, regardless of their stance on any other issue. People are very sensitive to contempt. People can sense very keenly when they are viewed primarily as problems.

This experience helped me realize that my work was very different from established projects in the field. A focus on connection, a focus on compassion, was really needed. A focus on respect, and personhood: these soft qualities were not well represented in outreach and engagement. But the field wanted these qualities. People in the field wanted to do better, they cared about doing more, they really cared about finding ways to reach people. I could represent these qualities in my work. I began to hope that in some small way I would be able to help change the culture. It was becoming easier for me to understand the many barriers that kept under-represented groups out of participation in the sciences, and this culture of contempt surely was one.

This is not to say that I was some hayseed come to the party. I had participated in decades of acculturation. I could talk as pretty and white as anyone. This had been an important skill during my studies in Arizona, where a white mouth and classical music on the radio convinced the Immigration and Customs Enforcement (ICE) to let me go twice, before even asking for my papers. I did always carry documentation, to be safe. I wore pretty jewels like the other ladies. Had gone hungry rather than sell them in the bad time, preserved my diamond earrings as crucial cultural markers. And I wore good clothing. Before the meeting, I found a Diane von Furstenberg in my size, silk, discounted to $40 from $500. My delight at this find knew no bounds.

I won't lie. I love diamonds, and I love silk dresses, and I would vastly prefer to have them than not to have them. They, and many other things I enjoyed at this time, like safe, quality childcare, a comfortable home, a reliable car, and good and regular food, were privileges that many people did not enjoy in my country at this time. Was it wrong for me to enjoy them? Did the simple possession of these things equate to contempt?

I don't think so, and I don't think equating privilege with contempt does anyone any favors. One can be a poor asshole as easily as a rich ass-hole. There are people with nothing who are nasty in their hearts and there are billionaires who are pure negatives. The problem was not privi-lege. The problem was decency, morality. Every cultural group has preju-dice and blindness.

I had always tried to keep a moral focus in my work. The president of my undergraduate institution, Illinois Wesleyan University, urged us to "do well, and to do good". I took those words deeply to heart. And in the beginning of 2017 I redoubled my commitment, and continued to explore in new ways, which were critical of my own culture, and the culture of my own institutions, to reach out to people as my fellow Americans.

In February, I learned that I would have more opportunities to do that than expected. The NSTA election results were in, and I had won. In June I would be installed as the new Division Director of Research in Science Education for the NSTA, a fairly powerful national organization that had significant policy influence, as well as direct impacts on tens of thousands of teachers in the United States and Canada.

This new elected position was awesome. For one thing, it meant that I had a voting seat on the board, which was clearly both powerful and cool. For another thing, it meant that I would meet many influential and interesting people. And then there was the most exciting part: that I would be in charge of the committee that recommended current research in science education to a very significant portion of North American science teachers. I would become a gatekeeper in determining what pedagogical research became practice. I would be able to keep expensive, poorly tested methods from national promotion, and work to give exposure to cost-effective, readily implementable methods.

As a person who was not climbing the tenure ladder, who did not need to publish lest I perish, I had unusual perspective and freedom on this issue. I could only imagine the political quagmire in which many people might find themselves once elected to this position. I was happy and grateful to be free to make decisions based on sound advice and my own best judgment.

The election results also placed me on the board of the National Association of Research in Science Teaching (NARST), though in a non-voting position. I knew from my experiences in academia that NARST could be a fairly competitive place. NARST was a place where conventional academic careers in my field were made, or broken. Again, what a wonderful place to have the chance to be truly an independent!

My work in Iowa was also developing nicely. My longitudinal dataset had finally grown to a point where I could do meaningful statistical analysis on at least one of my study sites. The first place where I could slap a year's worth of data together was Cedar Rapids, Iowa. We were able to show significant changes in scientific literacy.

Not only was the data significant, the change was really big. A twenty-four percent increase in the average score? What could explain that? Over the course of a year's study, the bottom had completely fallen out of the pool.

When looking at the numbers, the change was not so much because the average score itself was tremendously higher, but because the distribution had shifted dramatically to the right. We were no longer getting back surveys with the same kinds of very low scores we saw at the beginning of the study period, where people appeared to have had quite limited exposure to any kind of scientific information.

Was this change caused by the presence of our programming, or simply correlated to it? I didn't have enough evidence to say one way or the other. While we were collecting longitudinal data at several sites where our organization had a presence, we weren't collecting data in any communities where we weren't active. The logistics of getting people to take the survey apart from the treatment—how would you do it? And by this point we had spread to every area within reasonable driving distance. The effort it would take to regularly drive to a new community, more than an hour away, bust into community events similar to the ones in which we were currently participating, convince people to take surveys completely unconnected to anything benefiting them, and do it again every couple of months—it was just unreasonable. And we didn't have the funds to pay someone else to do it, or to pay a community to do it, and that would have injected new ethical problems into the mix anyway.

It was frustrating not to have a control group, but not unheard of in a longitudinal study of human subjects, exactly because of these ethical limitations. Our study was under IRB auspices, so it had to be conducted according to high ethical standards. In human subjects research, this sometimes if not often means that you cannot engage in exactly optimal data collection. It's not ethical to cause people harm, and treatment assignment can both actively and passively cause harm.

Active harm is easy to envision. Let's say you were interested in studying the impact of alcohol on domestic violence. You couldn't get couples with violent histories drunk with the intention of watching them slap each other around. Nor could you, for example, ethically give a bunch of people cigarettes under controlled conditions and see what factors contributed to the development of nicotine dependence. It's also morally murky to deny treatment. Let's say you had some burn patients. It would be obviously wrong to give some a new medication and leave others totally untreated.

Similarly, one can make an argument that education is a human need. That to withhold education when one could provide it is doing at the very least passive harm to a person or community. We see children in our society, as having both legal and moral rights to education.

Based on this ethical perspective, and our IRB restrictions, I would not be obtaining a control group any time soon. But maybe a year from now I'd have enough longitudinal data to show consistent improvement trends at sites where we became active. That would be something.

As March came to a close, multiple data sets seemed more and more like a hope worth having. All of the leaders had received their kits. And while not all of them had yet actually gone out on the ground, some of them had. Hundreds of people were being exposed to the materials we'd developed in Iowa, now across ten states. The network was small and fragile but it existed now and it hadn't existed before. I was doing something, which was better than nothing. And we had many people supporting us; we were receiving thousands of dollars in donations.

I just had to hold it together.

April–June 2017

Goodbye, Iowa

I had been terribly busy in the first quarter of 2017, managing the national expansion. It helped that the Iowa expansion was off my plate, as NCSE had hired Brian Pinney in Des Moines at my recommendation, starting in January. As the second quarter began, it was clear that things were going to continue to smooth out. Claire Adrain-Tucci had been hired to work part time with NCSE at my recommendation last fall and was now being promoted to full-time work, reporting to me. She would be taking over the day-to-day management of the national expansion, the assembly and delivery of kits, and the continual coaching and assistance the volunteer leaders required to help them succeed.

Putting all of this out of my hands was very frightening. Especially as it coincided with the move, I was concerned as usual that all of my life was a vicious trick and I was sure to be fired. Many people would have seen my situation very favorably. I now had three paid staff members reporting to me: a paid intern, a part-time staff member, and a full-time staff member. I'd have a fourth paid staff member, another intern, coming on that summer as a result of supplemental NSF funding through a partner lab. Another director-level staff member was being hired to do teacher support and further develop our outreach programs, and I was closely involved in the hiring process.

My job was becoming more managerial. I would no longer be on the ground at all. In the move to California, we would actually be living quite far out of Oakland, perhaps three hours from the office. We would be moving to the Sierra Nevada Mountains, where some of my most beloved family members were established. I would go into the office two days a week, spending the night in the city, and be with my family six nights a week.

I would be traveling often for my roles with NSTA and NARST. Ann did not want me doing any community organizing in my new area at all. I would work on strategic development and network with potential donors while I was in the office, and when I was not in the office I would continue to manage my staff at a distance and work on my writing. It would be time for me to start producing academic-quality publications based on my academic-quality work through the UIowa IRB.

That was what Ann wanted me to do, to do the higher-level work that only I could do, while it certainly did appear that I had enough of a structure in place to allow other people to carry on the work that I had developed in regard to outreach. How strange it was to say goodbye. I was ready to move on. I had to move on. I had too much other work to do with writing grants and my new responsibilities to be able to keep doing the ground work and management. But what a thing I had managed to build. It was a strange experience to step back and look at it.

My friends at the university found space for all of my gear. It took me quite a few trips to get everything over to the new lab. It did fill a whole room. My house was so curiously empty afterward. And my friends at the university also arranged to pay for my travel to the Netherlands that August, where there was a substantial gathering of evolutionary biologists, and where my collaboration with UIowa would be presented as part of a panel on the future of research funding.

These were very nice presents with quite substantial cash value. And very kind; I felt very appreciated. And I wondered if anything I had helped to build would actually hold together at all when I was gone or if it would all fall apart, and how sad that would be, because I did weep and weep as I packed up all of the things that were to go in my new lab where I would never work. Uncovering the many layers of supplies, it was so striking to get back to the beginning. Two years ago.

In two years, I had really built something. The SBC program was due to reach well over one hundred thousand people in 2017. It had contributed to the growing success and positive profile of our larger organization. In the first quarter of 2017, our donations were up substantially from the first quarter of 2016.

In 2017, the SBC itself had already raised more cash donations by May than we had in all of 2016. We had also already received a more than comparable to 2016 level of in-kind and UIowa staff support-based donations. Additionally, we had been invited to apply for state-level funding in Iowa by the Iowa Department of Natural Resources, which had the potential to pay five-figure levels of staff support.

Things were clearly growing in a very positive direction. I had some anxiety, perhaps even significant anxiety, that the program would not continue to grow if I were not constantly shepherding it. While I appreciated my staff and the work they did, I knew they were not as neurotic and obsessive as I was in regard to the work. They sometimes made mistakes or forgot to do things. Like people do. Like normal people, who seem to live without such terrible fear.

It was time for me to let go, develop the kind of baseline assumptions other people had. That things were going to be okay. That the world included something like stability.

At my quarterly review, Ann told me to spend the next quarter not working so hard. She suggested that, with the wedding, the move, and all those accompanying transitions, I make more of an effort to not go insane and less of an effort to expand the programs. She said that it was important I not burn out, that it would be a problem were I to do so.

So I supposed it was time to take that assignment. To be a bit less frantic. Go on more walks. Manage my staff, spend some time writing, give myself time to think.

Everything was fine. Over the past two years, I had accomplished many things. I had managed to expand access to scientific information to tens of thousands of Americans in what seemed likely to be a highly scalable fashion. I had found a way to measurably and significantly increase the scientific literacy of adults in communities. I was even finding some evidence for opinion shifts, that this increase in literacy inclined people to be more concerned about climate change (though not other scientific issues) and more interested in finding solutions.

Through my work at a nonprofit, I had found a way to apply my research more powerfully than I would have been able to in a traditional academic career. The way I had learned to talk to people so they could access scientific information—and so they could feel a sense of agency in regard to that information—was definitely working outside of a classroom setting. It was allowing us at NCSE to reach everyone, to make a real difference in the American landscape.

Of course, I had not single-handedly solved our country's problems, not even in a tiny corner of a small field. But I had done something, and now I had a seat at the table that helped determine policy in this field. I had a chance to do more.

And to do that, it was time to be okay.

CONCLUSION

Problems of access and power in science education have been studied extensively by a number of research communities. Community-level interventions, however, are not a typical form of educational research. Through work with a nonprofit, I was able to apply my research to a community-level intervention. While working with the socially divisive issues of climate change and evolution, my organization's programming correlated with real increases in community science literacy. But these literacy measures only tell part of the story. People enjoyed science education when it was presented in a fun, respectful way. They enjoyed the fruits of applied research tremendously. And it turned out that contentious topics were topics they found interesting, topics that they could accept or even care about, when these topics were presented without judgment.

The national expansion worked. By the end of 2017 we had a stable national network. We exceeded our goal, working with about one hundred and thirty thousand people over ten states. And that was with losing the Iowa State Fair in 2017, though we were accepted again for exhibition in 2018.

Solid research on how to talk with people about science was a big contributor to the success of this program. So was hard work. But the real driving force behind the success of this program was terrible need, terrible desperation.

We hear more and more stories about high-powered women in the workplace. We hear the stories of women who do great things, start great companies, and achieve enormous success. Many of these stories, and many of these women, are very admirable. But I could not see myself in any of these stories, which are most typically about women who do not have children, or who have supportive spouses, or who have more financial privilege than I had in my difficult time.

These stories were impossible for me to identify with, because they combine to present a context where female success is a matter of choice. We are encouraged to lean in, to develop our careers instead of stepping back to take care of family priorities. We are encouraged to develop careers for personal fulfillment. Those stories were not my story.

Many women work, work hard, and do amazing things from a place other than choice. They work because they need to support themselves and support their families. They work from places of suffering and pain. They work while experiencing domestic violence. They work through significant medical challenges.

I did the work described here while going through a divorce in a state where domestic abuse laws did not sufficiently protect myself or my children. In many American states my life would have wound up differently. I would have had adequate protection from my abusive spouse. If I had been protected, I would have been able to take a more moderate job for less money. If my child support was ensured by the state, as it is in some other countries, I would not have needed to fight for a promotion that advanced my career when my ex-husband abandoned his children.

I am a person of privilege. When I faced my big ordeal, I was not working full time because I had stepped back from my career to focus on my family responsibilities. There was a point where I had that economic choice. I had some financial resources to fall back on, some jewelry to sell. I had the enormous privilege of a truly excellent education, the vast majority of which was paid for by the American people due to federal support for STEM education. And I had a supportive community, a great gift.

Despite these privileges, my development of the SBC program was not a choice. Although I found the work fulfilling, fulfillment was a side effect rather than a driving factor. I put in so much work, pulled out all the stops, hurt my health for the work, because I needed the work to save my children.

I wanted to write fairly frankly about my experience doing this because the cultural narrative about women's work and choice is missing this experience. It is missing many of the ugly sides of the experiences American women have. Work can be fulfilling but it can also be a protection. Work can be a shield. Without my education I would not have been able to do the kind of work I did. I would have remained enslaved by my abuser, by the shackles of a state that frequently gives abusive men custody of their children. My children would have remained vulnerable.

A good science education should be the right of every American child. In a future where science education has a major impact on quality future career opportunities, science has a particularly strong ability to provide job opportunities and the possibility of upward mobility.

When your daughters ask you questions about women in the workplace, don't just tell them that work is fun, or work is a choice, or work is fulfilling. Acknowledge the reality: that a good education, and the possibility of a good job, is the most powerful thing a girl can have to help her leave unfulfilling, bad, or even abusive relationships when she is a grown woman. A good education lets a girl take care of herself. A good education gives women choices.

When we fight for educational access in the sciences we don't fight for pleasure. We don't fight for the frosting on the cake of an education. We fight for jobs, for meaningful choices, for safety and security.

If all kids could get the science education they deserve, it would be one small step toward liberation and dignity for all people. That is worth fighting for. Thank you for listening to how such education helped to give me safety, liberty, and dignity through the consideration of linguistic and cultural measures.

NOTES

1. Schoerning, E. (2014) The effect of plain-English vocabulary on student achievement and classroom culture in college science instruction. International Journal of Science and Mathematics Education 12(2): 307–327.
2. Richter, E. (2011) The Effect of Vocabulary on Introductory Microbiology Instruction. Arizona State University press, Tempe AZ.
3. Schoerning, E., & Hand, B. (2013) The discourse of argumentation. Mevlana International Journal of Education 2(3): 43–54.
4. Schoerning, E., & Hand, B. (2015) Language, access and power in the elementary science classroom. Science Education 99(2): 238–259.
5. I'll allow the organization to serve as a source on its own history. https://ncse.com/about/history
6. Plutzer, E., & Berkman, M. (2008) Trends: Evolution, creationism, and the teaching of human origins in schools. Public Opin Q 72(3): 540–553.
7. Plutzer, E., McCaffery, M., Hannah, A., Rosenau, J., Berbeco, M., & Reid, A. (2016) Climate confusion among US teachers. Science 351(6274): 664–665.
8. Iowa Pathways, by Iowa Public Television. http://site.iptv.org/iowapathways/timeline/1890--1899
9. Arne Duncan. (2011) Remarks by US Secretary of Education Arne Duncan to Iowa Education Summit https://www.ed.gov/news/speeches/iowas-wake-call
10. Laffer, A. B., Williams, J., & Moore, S. (2016) Rich States, Poor States, 9th Edition. ALEC center for state fiscal reform.

11. Hanushek, E., Peterson, P. E., & Woessman, L. (2012) Achievement Growth: International and US State Trends in Student Performance. Harvard's Program on Education Policy and Governance & Education Next.

12. Krob, G. (2016) Iowa's Changing Demographics. iowadatacenter.org. https://www.legis.iowa.gov/docs/publications/SI/794317.pdf

13. Next Generation Science Standards: Iowa. http://www.nextgenscience. org/iowa

14. Iowa Department of Public Health. (2015) 2014 Vital Statistics of Iowa. https://idph.iowa.gov/Portals/1/userfiles/68/HealthStats/vital_ stats_2014.pdf

15. CDC National Vital Statistics System: Marriage and Divorce Trends. https://www.cdc.gov/nchs/nvss/marriage_divorce_tables.htm

16. Krebs, V., & Holley, J. (2004) Building sustainable communities through social network development. The Nonprofit Quarterly 11: 46–53.

17. Krebs, V. (2008) Social capital: the key to success for the twenty-first century organization. IHRIM journal 12(5): 38–42.

18. Irvine, J. T. (1985) Status and style in language. Annual Review of Anthropology 14: 557–581.

19. Irvine, J. T. (1988) Ideologies of honorific language. Pragmatics 2(3): 251–262.

20. Schoerning, E., & Hand, B. (2015) Language, access and power in the elementary science classroom. Science Education 99(2): 238–259.

21. Matzke, N. (2016) The evolution of antievolution policies after *Kitzmiller* v *Dover*. Science 351(6268): 28–30.

22. Berkman, M., & Plutzer, E. (2010) Evolution, Creationism, and the Battle to Control America's Schools. Public Opinion Quarterly 72(3): 540–553.

23. Leiserowitz, A., Maibach, E., Roser-Renouf, C., Rosenthal, S., & Cutler, M. (2017) Climate change in the American mind: November 2016. Yale University and George Mason University. New Haven, CT: Yale Program on Climate Change Communication.

24. Ryan, M. (2016) Iowa's test score drop has officials asking why. http:// www.desmoinesregister.com/story/news/education/2016/10/ 12/iowas-test-score-drop-has-officials-asking-why/91569042/. Accessed 8/2/2017.

25. Munshi, N. (2016) Is gender neutrality the way to shut the stem gap? Financial Times. https://www.ft.com/content/85d23fde-d97d-11e5-a72f-1e7744c66818. Accessed 8/2/2017.

26. Calderwood, I. (2016) Daily Mail Online. http://www.dailymail.co.uk/ news/article-3466240/Anger-competition-aimed-getting-girls-interested-science-won-BOY.html. Accessed 8/2/2017.

27. After our success in the summer of 2016, the museum staff applied for space for another evolution-themed exhibit in 2017, and were accepted. In 2018, they are teaming up with NCSE again to present an exhibit on Genetics and Evolution at the Iowa State fair.

28. Schoerning, E. (2017) Community science literacy: Engagement and learning with conflict. Journal of Science Communication. Under revision.

29. Southern Poverty Law Center. (2016) Ten Days After: Harassment and Intimidation in the Aftermath of the Election. https://www.splcenter. org/20161129/ten-days-after-harassment-and-intimidation-aftermath-election. Accessed 3/15/2017.

REFERENCES

AAAS. (2014). *What we know.* www.whatweknow.aaas.org. Accessed 22 Jan 2018.

Abd-El-Khalick, F., et al. (2017). A longitudinal analysis of the extent and manner of representations of nature of science in US high school biology and physics textbooks. *Journal of Research in Science Teaching, 54*(1), 82–120.

Abi-El-Mona, I., & Abd-El-Khalick, F. (2006). Argumentative discourse in a chemistry high school classroom: An exploratory study. *School Science and Mathematics, 106*(8), 349–361.

Akkus, R., Gunel, M., & Hand, B. (2007). Comparing an inquiry-based approach known as the science writing heuristic to traditional science teaching practices: Are there differences? *International Journal of Science Education, 29*(14), 1745–1765.

Al-Azm, S. J. (2007). Islam and the science-religion debates in modern times. *European Review, 15*(3), 283–229.

Baker, J. (2012). Public perceptions of incompatibility between "science and religion". *Public Understanding of Science, 21*(3), 340–353.

Bandura, A. (2006). Toward a psychology of human agency. *Perspectives on Psychological Science, 7*(2), 164–180.

Bartos, S. (2014). Teacher's knowledge structures for nature of science and scientific inquiry: Conceptions and classroom practice. *Journal of Research in Science Teaching, 51*(9), 1150–1184.

Berkman, M., & Plutzer, E. (2010). Evolution, creationism, and the battle to control America's schools. *Public Opinion Quarterly, 72*(3), 540–553.

Bolger, D., & Ecklund, E. (2017). Whose authority? Perceptions of science education in Black and Latino churches. *Review of Religious Research.* https://doi.org/10.1007/s13644-017-0313-6.

© The Author(s) 2018   147
E. Schoerning, *Science Culture, Language, and Education in America,* https://doi.org/10.1057/978-1-349-95813-9

Boyd, M., & Rubin, D. (2006). How contingent questioning promotes extended student talk: A function of display questions. *Journal of Literacy Research, 38*(2), 141–159.

Branch, G. (2010). https://ncse.com/about/history. Accessed 30 Jan 2017.

Branch, G. (2014). Polling confidence in science. http://ncse.com/news/2014/04/polling-confidence-science-0015543. Accessed 15 Mar 2017.

Branch, G. (2015). Views on evolution amongst the public and scientists. http://ncse.com/news/2015/01/views-evolution-among-public-scientists-0016160. Accessed 15 Mar 2017.

Brown, B. (2001). Ch. 17, Orthodox Judaism. In J. Neusner & A. J. Avery-Peck (Eds.), *The Blackwell reader in Judaism* (p. 255). Oxford: Blackwell.

Calderwood, I. (2016). *Daily Mail* (Online). http://www.dailymail.co.uk/news/article-3466240/Anger-competition-aimed-getting-girls-interested-science-won-BOY.html. Accessed 2 Aug 2017.

Cavagnetto, A. R. (2010). Argument to foster scientific literacy: A review of argument interventions in K–12 science contexts. *Review of Educational Research, 80*(3), 336–371.

CDC National Vital Statistics System: Marriage and divorce trends. https://www.cdc.gov/nchs/nvss/marriage_divorce_tables.htm

Chen, Y.-C., Hand, B., & McDowell, L. (2013). The effects of writing-to-learn activities on elementary students' conceptual understanding: Learning about force and motion through writing to older peers. *Science Education, 97*(5), 745–771.

Choi, I., Nisbett, R. E., & Smith, E. E. (1997). Culture, category salience, and inductive reasoning. *Cognition, 65*(1), 15–32.

Choi, A., Notebaert, A., Diaz, J., & Hand, B. (2010). Examining arguments generated by year 5, 7, and 10 students in science classrooms. *Research in Science Education, 40*(2), 149–169.

Cimino, R., & Smith, C. (2011). The new atheism and the formation of the imagined secularist community. *Journal of Media and Religion, 10*, 24–38.

Clement, P. (2015). Creationism, science, and religion: A survey of teacher's conceptions in 30 countries. *Procedia – Social and Behavioral Sciences, 167*, 279–287.

Cobb, S. (2000). Negotiation pedagogy: Learning to learn. *Negotiation Journal, 16*(4), 315–319.

Crawford, B. A., Zembal-Saul, C., Munford, D., & Friedrichsen, P. (2005). Confronting prospective teachers' ideas of evolution and scientific inquiry using technology and inquiry-based tasks. *Journal of Research in Science Teaching, 42*(6), 613–637.

Creedy, D., & Hand, B. (1994). The implementation of problem-based learning: Changing pedagogy in nurse education. *Journal of Advanced Nursing, 20*(4), 696–702.

Dawkins, R. (2006). *The God delusion*. London: Bantam Books.

Delpit, L., & Dowdy, J. K. (2002). *The skin that we speak: Thoughts on language and culture in the classroom*. New York: New Press.

Diaz-Rico, L. T., & Weed, K. Z. (2002). *The crosscultural, language, and academic development handbook*. London: Allyn & Bacon.

Dukes, E. F. (1996). *Resolving public conflict: Transforming community and governance*. Manchester: Manchester University Press.

Duncan, A. (2011). *Remarks by US secretary of education Arne Duncan to Iowa education summit*. https://www.ed.gov/news/speeches/iowas-wake-call

Duran, B. J. (1998). Language minority students in high school: The role of language in learning biology concepts. *Science Education, 82*(3), 311–341.

Duschl, R., Ellenbogan, K., & Erduran, S. (1999). *Promoting argumentation in middle school science classrooms: A project SEPIA evaluation*. Paper presented at the Annual Meeting of the National Association for Research in Science, Boston, MA.

Edis, T. (2009). Modern science and conservative Islam: An uneasy relationship. *Science & Education, 18*(8), 885–903.

Emirbayer, M., & Mische, A. (1998). What is agency? *American Journal of Sociology, 103*(4), 962–1023.

Evans, J., & Evans, M. (2008). Religion and science: Beyond the epistemological conflict narrative. *Annual Review of Sociology, 34*, 87–105.

Falk, J., & Needham, M. (2011). Measuring the impact of a science center on its community. *Journal of Research in Science Teaching, 40*(1), 1–12.

Fisher, R., Ury, W., & Patton, B. (1991). *Getting to Yes: Negotiating agreement without giving in* (2nd ed.). New York: Penguin Books.

Fiske, J. (1994). *Media matters: Everyday culture and political change*. Minneapolis: University of Minnesota Press.

Ford, M., & Forman, E. A. (2006). Redefining disciplinary learning in classroom contexts. *Review of Research in Education, 30*(1), 1–32.

Gee, J. P. (1988). Dracula, the vampire lestat, and TESOL. *TESOL Quarterly, 22*(2), 201–226.

Gee, J. P. (1990). *Social linguistics and literacies: Ideology in discourses* (2nd ed.). London: Falmer.

Gelman, S. (2004). Psychological essentialism in children. *Trends in Cognitive Sciences, 8*(9), 404–409.

Gigliotti, D., Chernin, P., Topping, J., Williams, P., & Melfi, T. (Producers), & Melfi, T. (Director). (2016). *Hidden figures* [Motion picture]. Los Angeles: 20th Century Fox.

Gironi, F. (2010). Turning a critical eye on science and religion: Theological assumptions and soteriological rhetoric. *Method and Theory in the Study of Religion, 22*, 37–67.

Goodman, J. F., Hoadland, J., Pierre-Toussaint, N., Rodriguez, C., & Sanabria, C. (2011). Working the crevices: Granting students authority in authoritarian schools. *American Journal of Education, 117*(3), 375–398.

Gorham, J. (1988). The relationship between verbal teacher immediacy behaviors and student learning. *Communication Education, 37*(1), 40–53.

Gunel, M., & Hand, B. (2007). Comparing an inquiry-based approach known as the Science Writing Heuristic to traditional science teaching practices: Are there differences? *International Journal of Science Education, 29*(14), 1745–1765.

Hamza, K., & Wickman, P. (2007). Describing and analyzing learning in action: An empirical study of the importance of misconceptions in science learning. *Science Education.* https://doi.org/10.1002/sce.20233.

Hand, B., Norton-Meier, L., Staker, J., & Bintz, J. (2009). *Negotiating science: The critical role of argument in student inquiry, grades 5–10.* Portsmouth: Heinemann.

Hanushek, E., Peterson, P. E., & Woessman, L. (2012). *Achievement growth: International and US state trends in student performance.* Harvard's Program on Education Policy and Governance & Education Next.

Hathcoat, J., & Habashi, J. (2013). Ontological forms of religious meaning and the conflict between science and religion. *Cultural Studies of Science Education, 8,* 367–388.

Heath, S. B. (1983). *Ways with words: Language, life, and work in communities and classrooms.* Bath: Pittman Press.

Hildebrand, G. M. (2001). Re/writing science from the margins. In A. C. Barton & M. D. Osborne (Eds.), *Teaching science in diverse settings: Marginalized discourses and classroom practice* (pp. 161–199). New York: P. Lang.

Hodson, D. (1999). Going beyond cultural pluralism: Science education for sociopolitical action. *Science Education, 83*(6), 775–796.

Holdren, J. (2013). *America COMPETES: Science and the U.S. economy.* S.HRG. 113–641.

Iowa Department of Public Health. (2015). *2014 vital statistics of Iowa.* https://idph.iowa.gov/Portals/1/userfiles/68/HealthStats/vital_stats_2014.pdf

Iowa Pathways, by Iowa Public Television. http://site.iptv.org/iowapathways/timeline/1890--1899

Irvine, J. T. (1985). Status and style in language. *Annual Review of Anthropology, 14,* 557–581.

Irvine, J. T. (1988). Ideologies of honorific language. *Pragmatics, 2*(3), 251–262.

Isaacs, W. H. (1993). Taking flight: Dialogue, collective thinking, and organizational learning. *Organizational Dynamics, 22*(2), 24–39.

Jasanoff, S. (2010). A new climate for society. *Theory, Culture, and Society, 27*(2–3), 233–253.

Jiménez-Aleixandre, M. P. (2007). Designing argumentation learning environments. *Contemporary Trends and Issues in Science Education, 35*(2), 91–115.

Kahan, D. (2017). 'Ordinary science intelligence': A science-comprehension measure for study of risk and science communication, with notes on evolution and climate change. *Journal of Risk Research, 20*(8), 995–1016. https://doi.org/1 0.1080/13669877.2016.1148067.

Kahan, D., et al. (2016). Culturally antagonistic memes and the Zika virus: An experimental test. *Journal of Risk Research, 20*, 1–40.

Keep, S. (2018, January 10). Personal communication, NCSE staff.

Kintisch, E. (2005). U.S. economy. Panel calls for more science funding to preserve U.S. prestige. *Science, 310*(5747), 423.

Koenigs, S. S., Fiedler, M. L., & Decharms, R. (1977). Teacher belief, classroom interaction and personal causation. *Journal of Applied Social Pyschology, 7*(2), 95–114.

Krebs, V. (2008). Social capital: The key to success for the 21st century organization. *IHRIM journal, 12*(5), 38–42.

Krebs, V., & Holley, J. (2004). Building sustainable communities through social network development. *The Nonprofit Quarterly, 11*, 46–53.

Krob, G. (2016). *Iowa's changing demographics.* iowadatacenter.org. https://www.legis.iowa.gov/docs/publications/SI/794317.pdf

Laffer, A., Williams, J., & Moore, S. (2016). *Rich states, poor states* (9th ed.). ALEC Center for State Fiscal Reform. Arlington, VA.

Laugerman, M., Fostvedt, L., Shelley, M., Baenziger, J., Gonwa-Reeves, C., Hand, B., et al. (2013, March 7–9). *Structural equation modeling of knowledge content improvement using inquiry based instruction.* Interactive poster presentation at the Spring 2013 Conference of the Society for Research on Educational Effectiveness, Washington, DC.

Lee, O. (1997). Diversity and equity for Asian American students in science education. *Science Education, 81*(6), 107–122.

Lee, O., & Fradd, S. H. (1996). Literacy skills in science learning among linguistically diverse students. *Science Education, 80*(6), 651–671.

Leiber, D. (1999). *Etz Hayim, Torah and commentary.* New York: The Rabbinical Assembly.

Leiserowitz, A., Maibach, E., Roser-Renouf, C., Rosenthal, S., & Cutler, M. (2017). *Climate change in the American mind: November 2016.* Yale University and George Mason University. New Haven: Yale Program on Climate Change Communication.

Lemke, J. (1990). *Talking science: Language, learning and values.* Norwood: Ablex.

Lewis, C., Enciso, P., & Moje, E. B. (2007). *Reframing sociocultural research on literacy: Identity, agency, and power.* Mahwah: Erlbaum.

Long, D. (2010). Scientists at play in the field of the Lord. *Cultural Studies of Science Education, 5*, 213–235.

MacPhearson, J., & Kelly, S. (2011). Creativity and positive schizotypy influence the conflict between science and religion. *Personality and Individual Differences, 50*(4), 446–450.

Martin, A. M., & Hand, B. (2009). Factors affecting the implementation of argument in the elementary science classroom. A longitudinal case study. *Research in Science Education, 39*, 17–38.

Matzke, N. (2016). The evolution of antievolution policies after *Kitzmiller* v *Dover. Science, 351*(6268), 28–30.

McDonald, C. (2010). The influence of explicit nature of science and argumentation instruction on preservice primary teachers' views of nature of science. *Journal of Research in Science Teaching, 47*(9), 1137–1164.

Moje, E. B., Collazo, T., Carrillo, R., & Marx, R. W. (2001). "Maestro, what is 'quality'?": Language, literacy, and discourse in project-based science. *Journal of Research in Science Teaching, 38*(4), 469–498.

Munshi, N. (2016). Is gender neutrality the way to shut the stem gap? *Financial Times.* https://www.ft.com/content/85d23fde-d97d-11e5-a72f-1e7744c66818. Accessed 2 Aug 2017.

National Academies of Science, Engineering and Medicine Gulf Research Program. (2018). http://www.nas.edu/gulf/grants/education-2018/index.htm?_ga=2.122715915.1897361745.1515019309-1835389935.1513714083. Accessed 23 Jan 2018.

National Academy of Engineering. (2014). *Advancing diversity in the US industrial science and engineering workforce: Summary of a workshop.* Washington, DC: The National Academies Press.

National Oceanic and Atmospheric Administration Grant Impact Page. (2018). http://www.noaa.gov/office-education/elp/impacts. Accessed 23 Jan 2018.

National Science Board. (2016). *Science and engineering indicators 2016.* Arlington: National Center for Science and Engineering Statistics.

Navid, E., & Einsiedel, E. (2012). Synthetic biology in the science café: What have we learned about public engagement? *Journal of Science Communication, 11*(04), A02.

Neuman, S. B. (1996). Children engaging in storybook reading: The influence of access to print resources, opportunity, and parental interaction. *Early Childhood Research Quarterly, 11*, 495–513.

Newman, J. H. (1973). In C. S. Dessain & T. Gornall (Eds.), *The letters and diaries of John Henry Newman* (Vol. XXIV, pp. 77–78). Oxford: Clarendon Press.

Next Generation Science Standards: Iowa. http://www.nextgenscience.org/iowa

NGSS Lead States. (2013). *Next generation science standards: For states, by states.* Washington, DC: The National Academies Press.

Norton, M., & Nohara, K. (2009). Science cafes. Cross-cultural adaptation and educational applications. *Journal of Science Communication, 08*(04), A01.

Norton-Meier, L., Hand, B., Hockenberry, L., & Wise, K. (2008). *Questions, claims, and evidence: The important place of argument in children's science writing.* Portsmouth: Heinemann.

Nystrand, M., Wu, L. L., Zeiser, S., & Long, D. A. (2003). Questions in time: Investigating the structure and dynamics of unfolding classroom discourse. *Discourse Processes, 35*(2), 135–198.

Oversby, J. (2015). Teachers' learning about climate change education. *Procedia – Social and Behavioral Sciences, 167,* 23–27.

Pelletier, L. G., Seguin-Levesque, C., & Legault, L. (2002). Pressure from above and pressure from below as determinants of teachers' motivation and teaching behaviors. *Journal of Educational Psychology, 94,* 186–196.

Peng, K., & Nisbett, R. E. (1999). Culture, dialectics, and reasoning about contradiction. *American Psychologist, 54*(9), 741–754.

Plutzer, E., & Berkman, M. (2008). Trends: Evolution, creationism, and the teaching of human origins in schools. *Public Opinion Quarterly, 72*(3), 540–553.

Plutzer, E., McCaffery, M., Hannah, A., Rosenau, J., Berbeco, M., & Reid, A. (2016). Climate confusion among US teachers. *Science, 351*(6274), 664–665.

Pobiner, B. (2016). Accepting, understanding, teaching, and learning (human) evolution: Obstacles and opportunities. *Yearbook of Physical Anthropology, 159,* S232–S274.

Preston, J., & Epley, N. (2008). Science and God: An automatic opposition between ultimate explanations. *Journal of Experimental Social Psychology, 45,* 238–241.

Purcell-Gates, V. (2007). *Cultural practices of literacy: Case studies of language, literacy, social practice, and power.* Mahwah: Erlbaum.

Radloff, J. (2016). On teaching the nature of science: Perspectives and resources. *Cultural Studies of Science Education, 11,* 527–538.

Raiffa, H. (1982). *The art and science of negotiation.* Cambridge, MA: Harvard University Press.

Rakow, S. J., & Bermudez, A. B. (1993). Science is "Ciencia": Meeting the needs of Hispanic American students. *Science Education, 77*(6), 669–683.

Reeve, J., & Tseng, C. M. (2011). Agency as a fourth aspect of students' engagement during learning activities. *Contemporary Educational Psychology, 36*(4), 257–267.

Reeve, J., Deci, E. L., & Ryan, R. M. (2004). Self-determination theory: A dialectical framework for understanding the sociocultural influences on student motivation. In D. McInerney & S. Van Etten (Eds.), *Research on sociocultural influences on motivation and learning: Big theories revisited* (Vol. 4, pp. 31–59). Greenwich: Information Age.

Reiss, M. (2010). Science and religion: Implications for science educators. *Cultural Studies of Science Education, 5,* 91–101.

Reznitskaya, A., & Gregory, M. (2013). Student thought and classroom language: Examining the mechanisms of change in dialogic teaching. *Educational Psychologist, 48*(2), 114–133.

Richter, E. (2011). *The effect of vocabulary on introductory microbiology instruction.* Tempe: Arizona State University Press.

Rocklov, J. (2016). Misconceptions of global catastrophe. *Nature, 532,* 317–318.

Rosenau, J. (2015). Evolution, the environment, and religion. http://ncse. com/blog/2015/05/evolution-environment-religion-0016359. Accessed 13 May 2017.

Rosenthal, J. W. (1993). Theory and practice: Science for undergraduates of limited English proficiency. *Journal of Science Education and Technology, 2*(2), 435–443.

Rossato, C. A. (2007). *Engaging Paulo Freire's pedagogy of possibility: From blind to transformative optimism.* New York: Rowman & Littlefield.

Ryan, M. (2016). Iowa's test score drop has officials asking why. http://www. desmoinesregister.com/story/news/education/2016/10/12/iowas-test-score-drop-has-officials-asking-why/91569042/. Accessed 2 Aug 2017.

Sacks, J. (2011). *The great partnership: Science, religion, and the search for meaning.* New York: Schocken Books.

Schein, E. H. (1993). On dialogue, culture, and organizational learning. *Organizational Dynamics, 22,* 40–51.

Schoerning, E. (2014). The effect of plain-English vocabulary on student achievement and classroom culture in college science instruction. *International Journal of Science and Mathematics Education, 12,* 307–327.

Schoerning, E. (2017, under revision). Community science literacy: Engagement and learning with conflict. *Journal of Science Communication.*

Schoerning, E., & Hand, B. (2012). Language formality, learning environments and student achievement. In *The future of learning: Proceedings of the 10th International Conference of the Learning Sciences (ICLS 2012)* (Vol. 2, pp. 154–156). Sydney: ISLS.

Schoerning, E., & Hand, B. (2013). The discourse of argumentation. *Mevlana International Journal of Education, 2*(3), 43–54.

Schoerning, E., & Hand, B. (2015). Language, access and power in the elementary science classroom. *Science Education, 99*(2), 238–259.

Schurko, A., Neiman, M., & Logsdon, J. (2009). Signs of sex: What we know and how we know it. *Trends in Ecology, 24*(4), 208–217.

Settlage, J. (2007). Prognosis for science misconceptions research. *Journal of Science Teacher Education, 18*(6), 975–800.

Sevinc, B. (2011). Investigation of primary students' motivation levels towards science learning. *Science Education International, 22*(3), 218–232.

Skinner, E. A., & Belmont, M. J. (1993). Motivation in the classroom: Reciprocal effects of teacher behavior and student engagement across the school year. *Journal of Educational Psychology, 85,* 571–581.

Smith, S. S., & Dixon, R. G. (1995). Literacy concepts of low- and middle-class four-year-olds entering preschool. *Journal of Educational Research, 88*(4), 243–253.

Southern Poverty Law Center. (2016). Ten days after: Harassment and intimidation in the aftermath of the election. https://www.splcenter.org/20161129/ten-days-after-harassment-and-intimidation-aftermath-election. Accessed 15 Mar 2017.

Stears, M. (2012). Exploring biology education students' responses to a course in evolution at a South African university: Implications for their roles as future teachers. *Journal of Biological Education, 46*(1), 12–19.

Susskind, L., McKearnan, S., & Thomas-Larmer, J. (1999). *The consensus building handbook: A comprehensive guide to reaching agreement.* Thousand Oaks: Sage.

The Science Network. (2011). The moon, the tides and why Neil DeGrasse Tyson is Colbert's God a conversation about communicating science. Accessible online: http://thesciencenetwork.org/programs/the-science-studio/neil-degrasse-tyson-2. Accessed 3 Jan 2018.

Thomas, C., et al. (2004). Extinction risk from climate change. *Nature, 427,* 145–148.

Tippett, C. (2009). Argumentation: The language of science. *Journal of Elementary Science Education, 21*(1), 17–25.

Tobin, K., & McRobbie, C. J. (1996). Cultural myths as constraints to the enacted science curriculum. *Science Education, 80,* 223–241.

Tuan, H.-L. (2005). The development of a questionnaire to measure students' motivation towards science learning. *International Journal of Science Education, 27*(6), 639–654.

United Nations. (2015). *Paris agreement, English language text.* http://unfccc.int/files/essential_background/convention/application/pdf/english_paris_agreement.pdf. Accessed online 22 Jan 2018.

Uscinski, J., Douglas, K., & Lewandowsky, S. (2017). Climate change conspiracy theories. *Climate Science.* https://doi.org/10.1093/acrefore/9780190228620.013.328.

Vygotsky, L. S. (1987). Thinking and speech. In R. W. Rieber & A. S. Carton (Eds.), *The collected works of L.S. Vygotsky: Problems of general psychology* (Vol. 1, pp. 53–92) (N. Minick, Trans.). New York: Plenum Press. (Original work published 1934).

Walls, L. (2012). Third grade African America students' views of the nature of science. *Journal of Research in Science Teaching, 49*(1), 1–37.

Walsh, J. (1904). Audubon, the naturalist. *American Catholic Historical Society, 15*(1), 8–21.

Zembal-Saul, C., Munford, D., Crawford, B., Friedrichsen, P., & Land, S. (2002). Scaffolding preservice science teachers' evidence-based arguments during an investigation of natural selection. *Research in Science Education, 32*(4), 437–463.

Zimmerman, G. (2018). The Clergy Letter Project. http://www.theclergyletterproject.org/. Accessed 4 Jan 2018.

Index[1]

[1] Note: Page numbers followed by 'n' refer to notes.

© The Author(s) 2018
E. Schoerning, *Science Culture, Language, and Education
in America*, https://doi.org/10.1057/978-1-349-95813-9

157